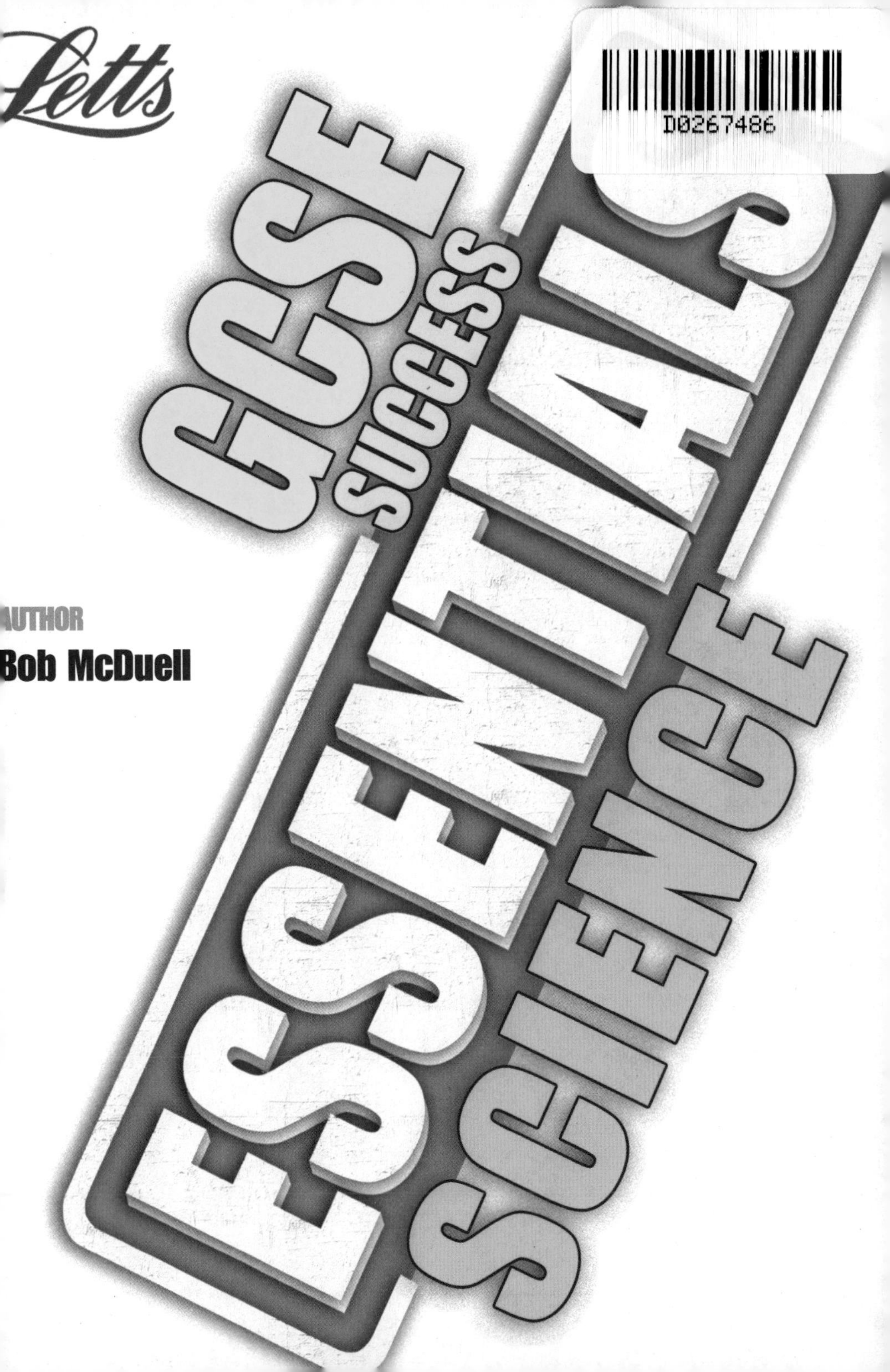
Letts
D0267486
GCSE
SUCCESS
ESSENTIALS
SCIENCE
AUTHOR
Bob McDuell

Contents

Introduction

Quick Reference

Tricky Topics

Tricky Topics

Coursework

Glossary

Published by Letts Educational
The Chiswick Centre
414 Chiswick High Road
London W4 5TF
tel: 020 89963333
fax: 020 87428390
e-mail: mail@lettsed.co.uk
website: www.letts-education.com

Letts Educational Limited is a division of Granada Learning Limited, part of Granada plc.

First published 2005

ISBN 1 84315 471 4

British Library Cataloguing in Publication Data

A catalogue record for this book is available from the British Library.

Acknowledgements
The publishers would like to thank the following for permission to reproduce photographs
p. 51 Pascal Goetgheluck/Science Photo Library,
p. 57 Martyn Chillmaid

Cover concept by Big Top

Commissioned by Cassandra Birmingham

Project managed by Julia Swales

Design and project management by Ken Vail Graphic Design, Cambridge

Printed in Italy.

Introduction

This book has been written as a GCSE revision product for students studying for Double Award and Single Award science courses. It also supplements the *GCSE Success Guides* and *GCSE Success Guides Question and Answers* for physics, chemistry and biology. Included are some useful tips and an emphasis on those areas that most students find difficult. There is also information to enable students to do better in their science coursework.You will find references to the *Success Guides* – Foundation (**F**) and Higher (**H**) – where you can look up more information, at the bottom of most pages.

The book has been split into several sections.

Essential information

The book is useful to keep at hand during science lessons, while doing your homework and while revising for your GCSE examination. The **Quick Reference** section contains the essential information that you will have to use throughout the course and also some useful facts that you need to know.

Tackling the tricky topics

Each year an examiners' report is written for each GCSE science specification. The report highlights areas where candidates did well in the examination and areas where they made mistakes. We have called these areas **Tricky Topics**. This book highlights several Tricky Topics and helps you master them. Many of them appear in Higher-tier questions, so being able to handle them will boost your confidence.

Get to grips with coursework

Since the coursework element of GCSE science counts for up to 20% of the total mark, it is important to take time over it and get as high a mark as possible. The section on **Coursework** will help you tackle this area. Science coursework is done and marked under four headings – Planning, Obtaining evidence, Analysing and Evaluating. The key to success is making a good plan based on real and correct science at the highest level. Material in the Tricky Topics section will also be helpful here.

Learn your lingo

To succeed really well, you need to understand and use the correct scientific vocabulary. The last section of this book is a **Glossary** that will help you when you come across unfamiliar scientific words in lessons or when doing your homework.

How to revise science

- Concentrate on the topics you need to work on most. However, you should not miss out topics as most of the specification will be examined in each set of exams.
- Here are a few tips to help you with your revision.

Planning

- Find out the dates of your science examinations.
- Make an exam and revision timetable.
- Look at a copy of the specification. You may be able to get a copy from your teacher. If not, specifications can be downloaded from the websites listed below. Go through and make a list of topics. Look at the list and decide honestly which ones you are happy with and which need more work. Concentrate on the latter. Deal with them one at a time.
- For each topic, make some brief notes (no more than one side of A4) of key points as you revise. Keep these notes because they will be helpful for last-minute revision.
- Try and write out these key points from memory. Check what you have written against your list. Have you missed anything out?

Revising

- Revise in short bursts of about 30 minutes, followed by a short break.
- Memorise any formulae you need to learn. Learning with a friend is easier and more fun.
- Try sample questions of all types. Make sure you practise Extended Writing questions: these are questions requiring long answers and candidates often have difficulty with them.

Taking the examination

- The night before the exam, try to have an early night. Exams are tiring and you need to be alert.
- Get all of your equipment – pens, pencil, ruler, calculator – ready the day before.
- Arrive in good time.
- Follow all the instructions on the exam paper. In the science exam, all the questions are usually compulsory, so when you find something you know you can do, start there. A strong start will give you confidence.

We hope you find this book useful during your coursework, revision and as you prepare for the exams.

Good luck!

Websites

www.aqa.org.uk
www.ccea.org.uk
www.edexcel.org.uk
www.orc.org.uk
www.wjec.co.uk

Command words

Each part of every question includes a word that tells you what you should do. It is called a **command word**. If you ignore the command word or do something other than what it says, you will lose marks.

Here are some of the main command words with suitable responses.

State – This requires a **brief answer** without any explanation. Sometimes the words 'Write down' or 'Give' might be used.

Define – Here a **precise meaning** of a word or phrase is expected.

Describe – This requires a more **detailed answer** with all the **relevant information** given in the **correct order**. This type of question is always worth several marks.

Outline – This requires the **description** of a process or event but only the **main details** have to be given.

Compare – This requires you to give an **account** of the main **similarities and differences** between two or more things.

The answer here needs **mathematical methods** to complete it. Show all your **working** and any relevant unit.

Here there is more than one possible answer. Any **sensible** answer will score marks.

Here you are expected to give both sides of an **argument** and possibly your **opinion**.

Here you should use your knowledge to **assess** the implications and limitations of the information given.

Your answer should be a **detailed account** including causes, reasons, etc. This type of question will always be worth a number of marks.

Use the information in the question to **continue a pattern or trend**.

The Periodic Table

As the elements were discovered, early chemists tried to find patterns amongst them, but they struggled to find links.

Group	I	II											III	IV	V	VI	VII	0
Period																		
1	**H** 1																	**He** 2
2	**Li** 3	**Be** 4											**B** 5	**C** 6	**N** 7	**O** 8	**F** 9	**Ne** 10
3	**Na** 11	**Mg** 12											**Al** 13	**Si** 14	**P** 15	**S** 16	**Cl** 17	**Ar** 18
4	**K** 19	**Ca** 20	**Sc** 21	**Ti** 22	**V** 23	**Cr** 24	**Mn** 25	**Fe** 26	**Co** 27	**Ni** 28	**Cu** 29	**Zn** 30	**Ga** 31	**Ge** 32	**As** 33	**Se** 34	**Br** 35	**Kr** 36
5	**Rb** 37	**Sr** 38	**Y** 39	**Zr** 40	**Nb** 41	**Mo** 42	**Tc** 43	**Ru** 44	**Rh** 45	**Pd** 46	**Ag** 47	**Cd** 48	**In** 49	**Sn** 50	**Sb** 51	**Te** 52	**I** 53	**Xe** 54
6	**Cs** 55	**Ba** 56	57 – 71*	**Hf** 72	**Ta** 73	**W** 74	**Re** 75	**Os** 76	**Ir** 77	**Pt** 78	**Au** 79	**Hg** 80	**Tl** 81	**Pb** 82	**Bi** 83	**Po** 84	**At** 85	**Rn** 86
7	**Fr** 87	**Ra** 88	89 – 103‡	**Rf** 104	**Db** 105	**Sg** 106	**Bh** 107	**Hs** 108	**Mt** 109	**Uun** 110	**Uuu** 111	**Uub** 112	**Uut** 113	**Uuq** 114	**Uup** 115	**Uuh** 116	**Uus** 117	**Uuo** 118

‡Lanthanides	**La** 57	**Ce** 58	**Pr** 59	**Nd** 60	**Pm** 61	**Sm** 62	**Eu** 63	**Gd** 64	**Tb** 65	**Dy** 66	**Ho** 67	**Er** 68	**Tm** 69	**Yb** 70	**Lu** 71
***Actinides**	**Ac** 89	**Th** 90	**Pa** 91	**U** 92	**Np** 93	**Pu** 94	**Am** 95	**Cm** 96	**Bk** 97	**Cf** 98	**Es** 99	**Fm** 100	**Md** 101	**No** 102	**Lr** 103

Note that elements 113, 115 and 117 are not yet known, but are included in the table to show their respective positions. Elements 114, 116 and 118 have only been reported recently.

Key:
- Non-metal
- Metalloid
- Metal
- Transitional
- Rare-earth element (Lanthanide) and radioactive rare-earth element (Actinide)
- Transactinide
- ☐ 'Missing' element

Common elements

Element	Symbol	Metal or non-metal	Melting point (°C)	Boiling point (°C)	Density (g/cm³)	Date of discovery
Hydrogen	H	non-metal	–259	–253	0.00008	1766
Helium	He	non-metal	–270	–269	0.00017	1868
Lithium	Li	metal	180	1330	0.53	1817
Carbon	C	non-metal	**	4200	2.2	*
Nitrogen	N	non-metal	–210	–196	0.00117	1772
Oxygen	O	non-metal	–219	–183	0.00132	1774
Fluorine	F	non-metal	–220	–188	0.0016	1886
Neon	Ne	non-metal	–249	–246	0.0008	1898
Sodium	Na	metal	98	890	0.97	1807
Magnesium	Mg	metal	650	1110	1.7	1808
Aluminium	Al	metal	660	2060	2.7	1825
Silicon	Si	non-metal	1410	2700	2.4	1825
Phosphorus	P	non-metal	44	280	1.8	1669
Sulphur	S	non-metal	119	444	2.1	*
Chlorine	Cl	non-metal	–101	–35	0.003	1774
Argon	Ar	non-metal	–189	–189	0.0017	1894
Potassium	K	metal	64	760	0.86	1807
Calcium	Ca	metal	850	1440	1.6	1808
Manganese	Mn	metal	1250	2000	7.4	1774
Iron	Fe	metal	1540	3000	7.9	*
Copper	Cu	metal	1080	2500	8.9	*
Zinc	Zn	metal	419	910	7.1	17th Century
Bromine	Br	non-metal	–7	58	3.1	1826
Krypton	Kr	non-metal	–157	–153	0.0035	1898
Silver	Ag	metal	961	2200	10.5	*
Iodine	I	non-metal	114	183	4.9	1811
Barium	Ba	metal	710	1600	3.5	1805
Gold	Au	metal	1060	2700	19.3	*
Mercury	Hg	metal	–39	357	13.6	*
Lead	Pb	metal	327	1744	11.3	*

** These elements have been known for thousands of years.   ** Does not melt*

Electron arrangement

Element	Atomic number	Mass number	Number of p	n	e	Electron arrangement
Hydrogen	1	1	1	0	1	1
Helium	2	4	2	2	2	2
Lithium	3	7	3	4	3	2,1
Beryllium	4	9	4	5	4	2,2
Boron	5	11	5	6	5	2,3
Carbon	6	12	6	6	6	2,4
Nitrogen	7	14	7	7	7	2,5
Oxygen	8	16	8	8	8	2,6
Fluorine	9	19	9	10	9	2,7
Neon	10	20	10	10	10	2,8
Sodium	11	23	11	12	11	2,8,1
Magnesium	12	24	12	12	12	2,8,2
Aluminium	13	27	13	14	13	2,8,3
Silicon	14	28	14	14	14	2,8,4
Phosphorus	15	31	15	16	15	2,8,5
Sulphur	16	32	16	16	16	2,8,6
Chlorine	17	35	17	18	17	2,8,7
Argon	18	40	18	22	18	2,8,8
Potassium	19	39	19	20	19	2,8,8,1
Calcium	20	40	20	20	20	2,8,8,2

Common compounds and their formulae

Name	Formula
Carbon dioxide	CO_2
Water	H_2O
Sulphuric acid	H_2SO_4
Hydrochloric acid	HCl
Nitric acid	HNO_3
Sodium hydroxide	NaOH
Potassium hydroxide	KOH
Calcium hydroxide	$Ca(OH)_2$
Sodium hydrogencarbonate (bicarbonate of soda)	$NaHCO_3$
Carbonic acid	H_2CO_3
Manganese(IV) oxide	MnO_2
Sodium chloride (salt)	NaCl
Sodium sulphate	Na_2SO_4
Sodium nitrate	$NaNO_3$
Iron(III) oxide	Fe_2O_3
Aluminium oxide	Al_2O_3
Magnesium oxide	MgO
Calcium carbonate	$CaCO_3$
Methane	CH_4
Sucrose (sugar)	$C_{12}H_{22}O_{11}$
Glucose	$C_6H_{12}O_6$
Fructose	$C_6H_{12}O_6$
Ethanol (alcohol)	C_2H_5OH

Common units and symbols

Quantity	Unit (symbol)	
mass	kilogram (kg) milligram (mg)	gram (g) microgram (μg)
length	micrometre (μm) centimetre (cm) kilometre (km)	millimetre (mm) metre (m)
volume	cubic metre (m^3) cubic decimetre (dm^3) cubic centimetre (cm^3) litre (l)	millilitre (ml)
time	second (s) hour (h)	minute (min) year (y)
temperature	degree Celsius (°C)	Kelvin (K)
chemical quantity	mole (mol)	
potential difference (voltage)	volt (V)	
current	ampere (A)	milliampere (mA)
resistance	ohm (Ω) megohm (MΩ)	kilohm (kΩ)
force	newton (N)	
energy (work)	joule (J) kilowatt-hour (kWh)	kilojoule (kJ)
power	watt (W)	kilowatt (kW)
density	kilogram per cubic metre (kg/m^3) gram per cubic centimetre (g/cm^3)	
concentration	mols per cubic decimetre (mol/dm^3)	

Physics equations

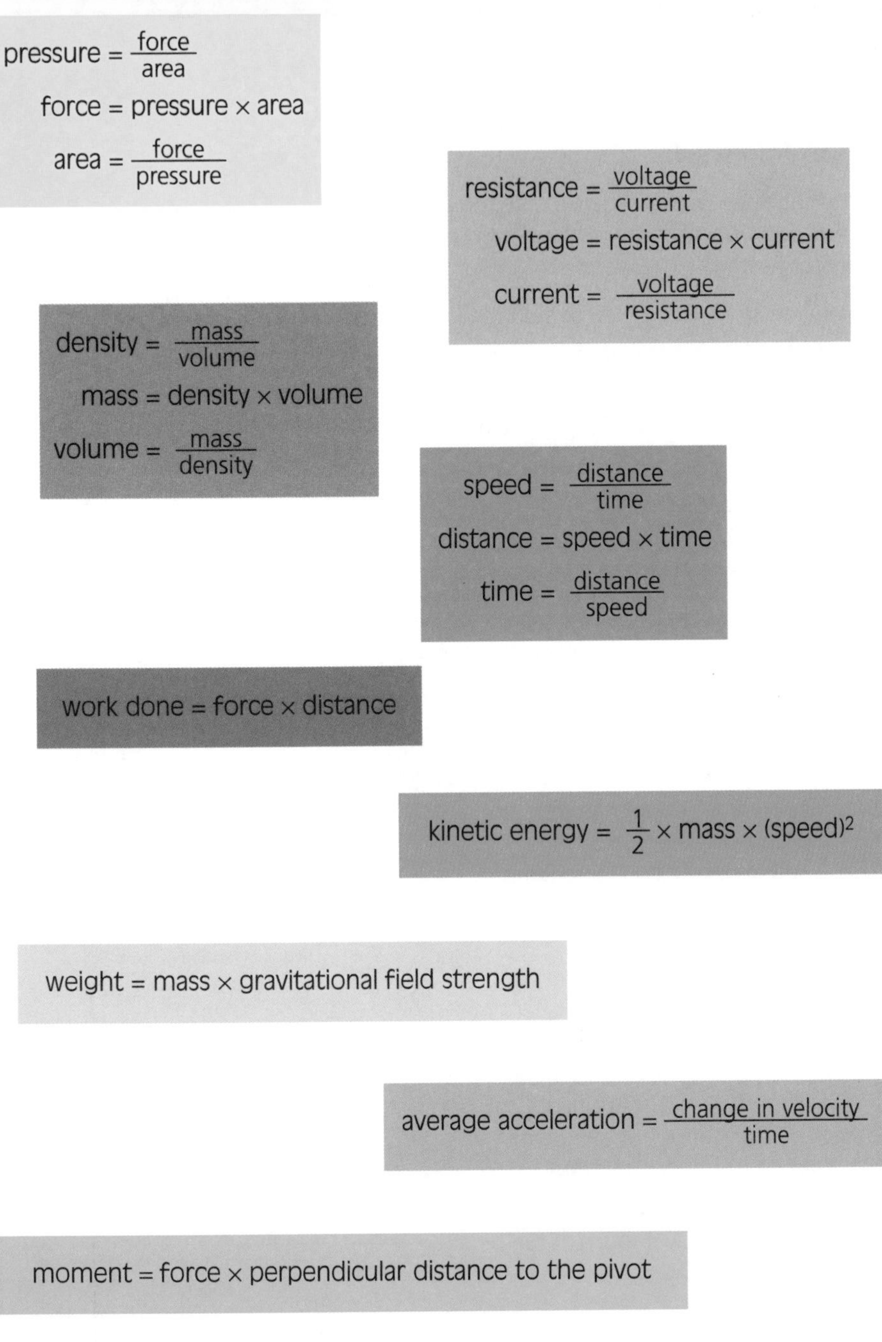

$$\text{pressure} = \frac{\text{force}}{\text{area}}$$

$$\text{force} = \text{pressure} \times \text{area}$$

$$\text{area} = \frac{\text{force}}{\text{pressure}}$$

$$\text{resistance} = \frac{\text{voltage}}{\text{current}}$$

$$\text{voltage} = \text{resistance} \times \text{current}$$

$$\text{current} = \frac{\text{voltage}}{\text{resistance}}$$

$$\text{density} = \frac{\text{mass}}{\text{volume}}$$

$$\text{mass} = \text{density} \times \text{volume}$$

$$\text{volume} = \frac{\text{mass}}{\text{density}}$$

$$\text{speed} = \frac{\text{distance}}{\text{time}}$$

$$\text{distance} = \text{speed} \times \text{time}$$

$$\text{time} = \frac{\text{distance}}{\text{speed}}$$

$$\text{work done} = \text{force} \times \text{distance}$$

$$\text{kinetic energy} = \frac{1}{2} \times \text{mass} \times (\text{speed})^2$$

$$\text{weight} = \text{mass} \times \text{gravitational field strength}$$

$$\text{average acceleration} = \frac{\text{change in velocity}}{\text{time}}$$

$$\text{moment} = \text{force} \times \text{perpendicular distance to the pivot}$$

Graphs

In many questions, both at Foundation and Higher tier, you might be expected to draw **graphs**. You will also probably draw graphs in Sc1 coursework.

At Foundation tier you will be given scales and axes and the plotting will be **straightforward**.

At Higher tier you will be expected to **choose** appropriate scales and axes, then **plot data** where some of the data is anomalous and, if appropriate, **draw** a line of best fit.

Here are some things you should remember when drawing graphs.

1. If you have a choice use 2mm graph paper.
2. The independent variable goes along the ***x*** (horizontal) axis and the dependent variable along the ***y*** axis (vertical). Label the axes and give units.
3. If you are given a grid, choose scales that fit more than half of the graph paper.
4. Make sure your scales are linear.
5. Plot the points carefully, either with a small cross or a dot with a circle around it.
6. If the values on the ***x*** axis are discrete, drawing a line of best fit is inappropriate. For example, Graph A opposite shows the number of males and females suffering from asthma during their lifetime. The results are available only every five years. A line of best fit is inappropriate and the points should be joined with straight lines.
7. In graph B, time is a continuous variable and so a graph of volume of gas collected against time should be drawn as a line of best fit. Use a sharp pencil to do this and miss out any points that do not match the trend.
8. Finally, give the graph a title. This is especially important for Sc1 coursework.

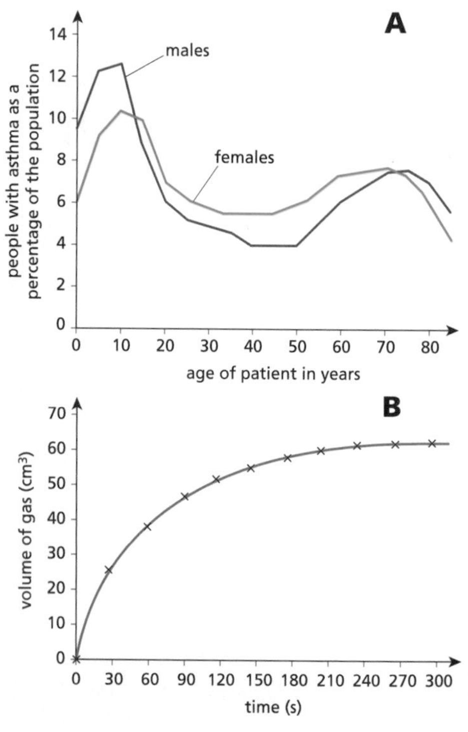

Calculations

On all science papers there will be chemical **calculations**. You can use a calculator on all papers.

In **physics** most of the calculations involve **recall** and use of certain scientific **equations** (see page 13). When attempting one of these calculations, first write down the relationship you are going to use. At Higher tier you are expected to **rearrange** these relationships but at Foundation tier just **substitute** values and work out an answer. In **chemistry** many of the calculations involve chemical **formulae** and **equations**. These calculations require ratio calculations. The calculations in **biology** are more **varied**.

Show your working

When you attempt a calculation always **show your working in full**. If a calculation is worth 3 marks and you get it correct without showing any working you will get 3 marks. However, if you make a mistake without showing your working, your work is worth 0 marks. If you show working the examiner can work back to see where you went wrong and will possibly award 1 or 2 marks.

Correct significant figures

Candidates using a calculator often write the answer straight from the calculator screen. So if 4.2g of a chemical are used and 3.3333333g are produced, the candidate should **correct** it to 3.3g, the same number of **significant figures** as in the question.

All GCSE papers have opportunities for you to be awarded marks for Quality of Written Communication. These marks are in addition to the marks awarded for the science itself.

Often these questions are identified by a statement explaining what aspect of Quality of Written Communication (QWC) is being tested. Alternatively there might be an icon next to the question.

Sometimes the additional marks are shown in the right-hand margin. So 3+1 means 3 marks for science plus 1 mark for QWC.

What is the examiner looking for?

Various things can be tested by Quality of Written Communication (QWC) and different questions provide different opportunities.

Using correct spelling, punctuation and grammar

You will be awarded an additional mark if you write your answers with correct sentences. These sentences must start with a capital letter, contain a verb and end with a full stop. If the mark is available, writing in bullet points or phrases might gain you full marks for science but not the QWC mark. Read your answer carefully after you have written it to make sure there are no simple mistakes in spelling, punctuation and grammar.

Writing a clear and ordered answer

Often candidates writing a longer answer, such as a description of an experiment, make perfectly sound scientific points but put them in an illogical order.

If the QWC mark is going to be awarded for a clear, ordered answer you will be given an additional mark if the Examiner judges you have put your answer in a logical order. Read your answer carefully and make sure that there are no gaps or inconsistencies in your answer. When revising, get used to writing longer answers by noting the points you want to make and then putting them into the most logical order.

Using correct scientific terminology

Candidates often fail to use scientific language or fail to use it correctly. If the mark is awarded for correct scientific terminology, the examiner is looking for the correct use and spelling of certain technical words.

Technical words such as **mitosis** or **oxidising agent** or **momentum** have very clear scientific meaning. If terms are misused the examiner will not award the mark.

It is a good idea to make a list of technical words as part of your revision. Go through your textbook or notes or the glossary on page 94 of this book. Alternatively, look at a scientific dictionary.

Make sure you can spell each word and understand exactly what it means. Try to read each word in a sentence to confirm how it is used.

Marks for QWC are awarded in a similar way at AS and A2 level. Each mark has the same value and it is important that you do not miss out on this relatively simple way to gain marks.

Solids, liquids and gases

Students wanting to achieve a grade higher than C need to have a clear understanding of the **arrangement and movement of particles** in a range of circumstances.

All substances can exist in the three states, **solid, liquid** and **gas**, depending upon conditions of temperature and pressure.

The diagram shows two-dimensional representations of the arrangement of particles in a solid, a liquid and a gas.

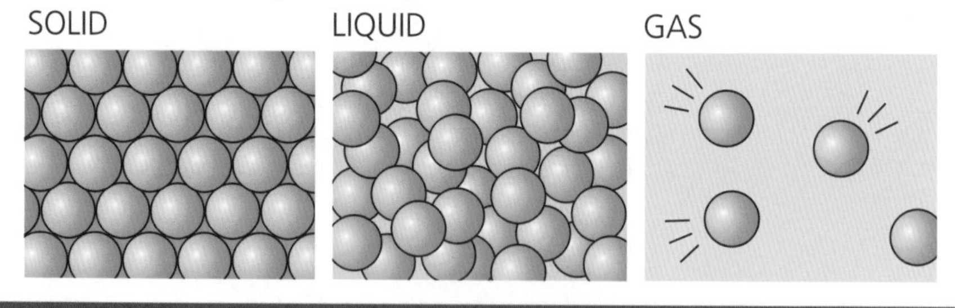

Key features in these diagrams

In a solid the particles are regularly arranged. This regular arrangement leads to a crystalline structure.

In a liquid the particles are still close together but there is no regular arrangement.

In a gas the particles are more spread out and again there is no regular pattern.

Students frequently have difficulty drawing these diagrams. This is because they:

- **either** draw the particles too large, so the arrangement is impossible to see
- **or** draw the particles too small so that it takes a long time to fill the space.

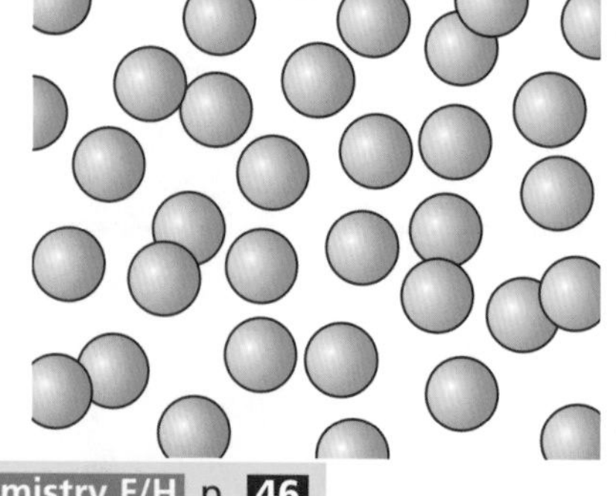

Many students draw a liquid as shown here. This representation would not explain the fact that liquids are not easily compressed. This diagram better represents a gas at high pressure.

For more help see **Success Guide Chemistry F/H** p. **46**

Movement of particles in solids, liquids and gases

- The particles in a solid only vibrate with little movement.
- In a liquid, the particles move more and are able to move throughout the liquid, but only very slowly because of the congestion of particles.
- Particles in a gas move rapidly in random motion.
- As temperature increases the particles in solids, liquids and gases move faster.

Changes of state

The diagram shows the relationships between the three states of matter.

If a solid is heated with a steady source of energy, the graph shows how the temperature changes.

The flat portions of the graph correspond to where changes of state take place. Here the energy received is used to move particles apart and give them more energy.

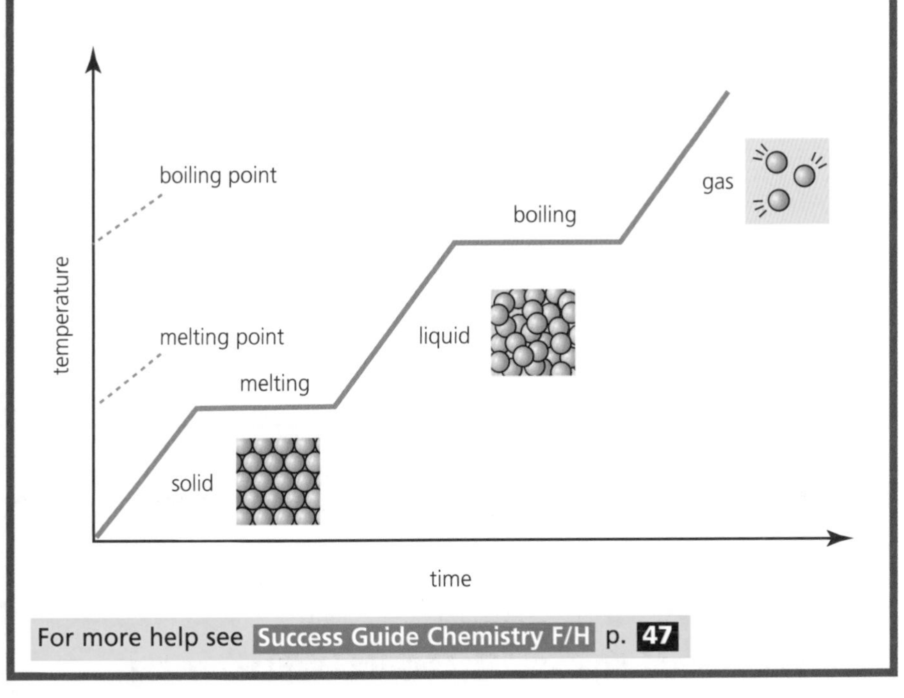

For more help see Success Guide Chemistry F/H p. 47

Diffusion

Diffusion is the spontaneous movement of particles from an area of **high concentration** to an area of **low concentration**. It is also the natural tendency of particles of matter to fill all available space.

Diffusion can be demonstrated using bromine gas.

A gas jar filled with brown bromine vapour is placed below a gas jar filled with air. In a few minutes both gas jars are filled with the same mixture of bromine and air.

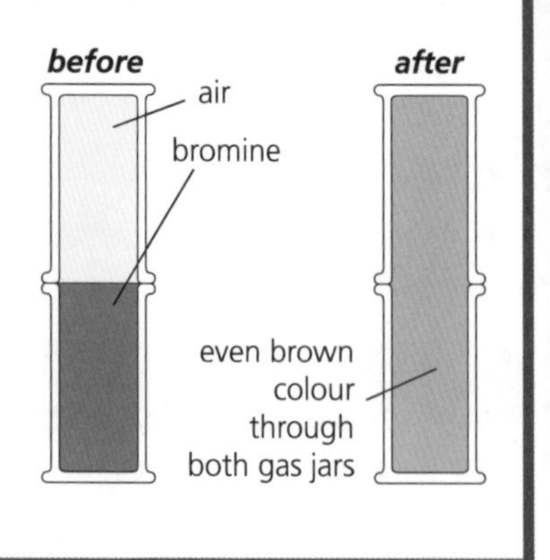

Diffusion in solids, liquids and gases

Diffusion takes place quickly with gases and slowly with liquids. This is because of the greater movement of particles in gases. The diagram shows the slow diffusion of potassium manganate(VII) in water without stirring.

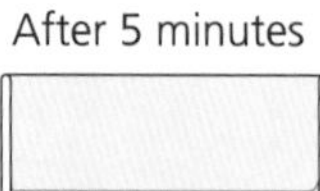
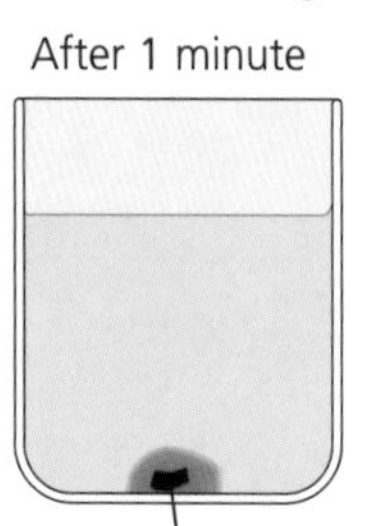
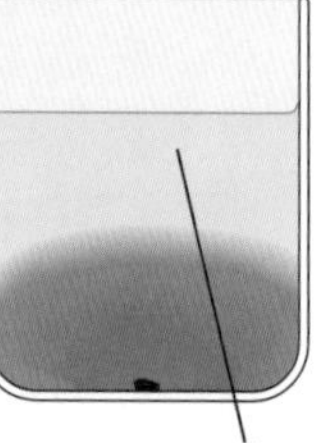

Diffusion can take place in solids but it is a very slow process taking many years.

Concentration gradient

One factor that determines how quickly diffusion takes place is the difference between the two concentrations at the start. The bigger the difference in concentration, the steeper the concentration gradient.

For more help see Success Guide Chemistry F p. 33/H p. 25

Rates of diffusion of gases

A second factor affecting the rate of diffusion is the mass of the particles involved.

A dry tube is clamped horizontally and pads of cotton wool soaked in ammonia solution and hydrochloric acid are put into opposite ends of the tube.

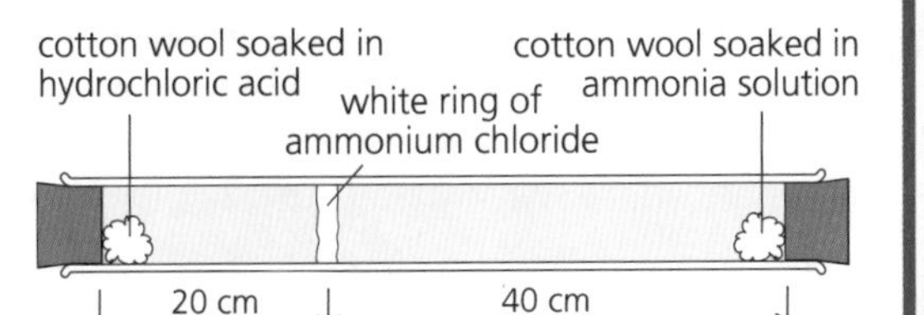

The particles of ammonia and hydrogen chloride move along the tube and a white solid, ammonium chloride, is formed when they meet. After about 5 minutes a white ring forms closer to the hydrochloric acid end of the tube. The lighter ammonia particles move faster than the heavier hydrogen chloride particles.

Diffusion in plant cells

Carbon dioxide, needed for photosynthesis (see pages 54–5), diffuses into the leaves through tiny holes called stomata in the underside of leaves. The products of photosynthesis diffuse out.

palisade cells
spongy cells
lower epidermis
O_2
CO_2
stomata
guard cell
carbon dioxide
water & oxygen

Diffusion in animal cells

Body cells need glucose and oxygen for respiration (see pages 56–7). These are carried in the blood and then diffuse through the capillary walls into cells. The products, carbon dioxide and water, diffuse out of the cell.

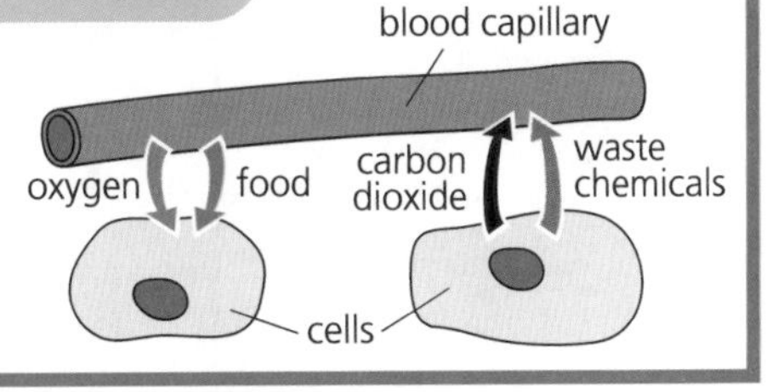

Active transport

Some examples in biological systems cannot be explained by simple diffusion.

For example, root hair cells absorb ions from the soil even when the concentration of ions is greater inside the cell than in the soil around. This is called **active transport** and needs energy from respiration to take place.

For more help see Success Guide Chemistry F p. 33/H p. 28

Osmosis

Osmosis is a special form of **diffusion**. It involves the diffusion of water through a **partially permeable membrane** from an area of high concentration (i.e. water or a dilute solution) to an area of low concentration.

In the diagram, water passes from dilute solution on the left through the semi-permeable membrane.

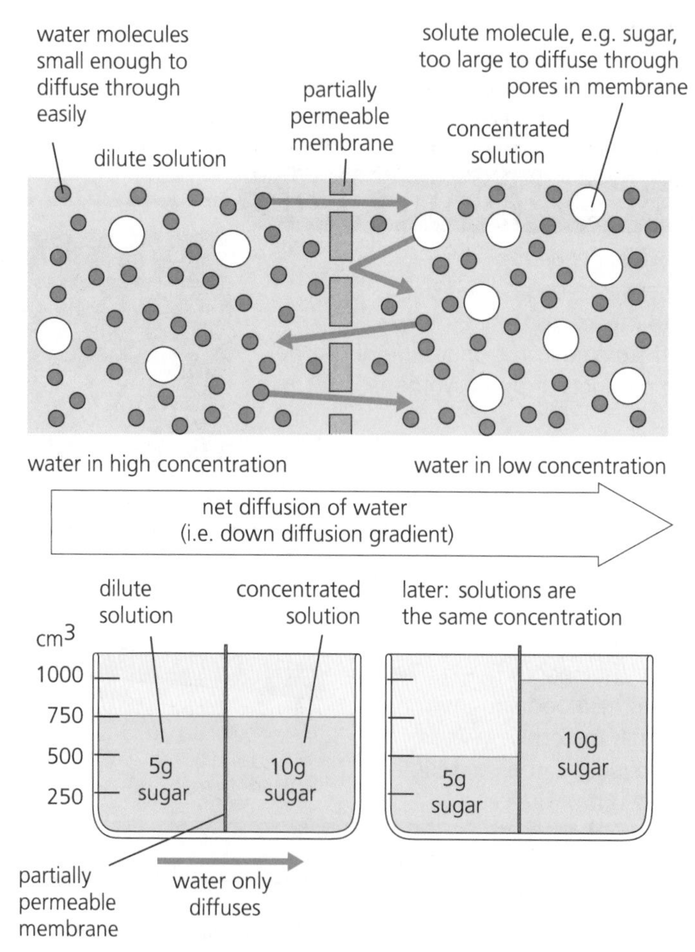

At the end of the process, water molecules pass from the left to the right at the same rate as they pass from the right to the left. An **equilibrium** is set up.

For more help see **Success Guide Biology F** p. **32**/**H** p. **29**

Osmosis using pieces of potato

The table compares what happens when pieces of tissue, cut from the same potato, are put into:

a) distilled water

b) concentrated sugar solution.

	a) distilled water	b) concentrated sugar solution
Appearance	Potato became hard, stiff and enlarged. It became **turgid**.	Potato became soft, floppy and shrank. It became **flaccid**.
Mass of sample	increases	decreases
Explanation	Water moves into the potato by osmosis, as the water concentration in surroundings is greater.	Water moves out of the potato by osmosis, as the water concentration in potato is greater.

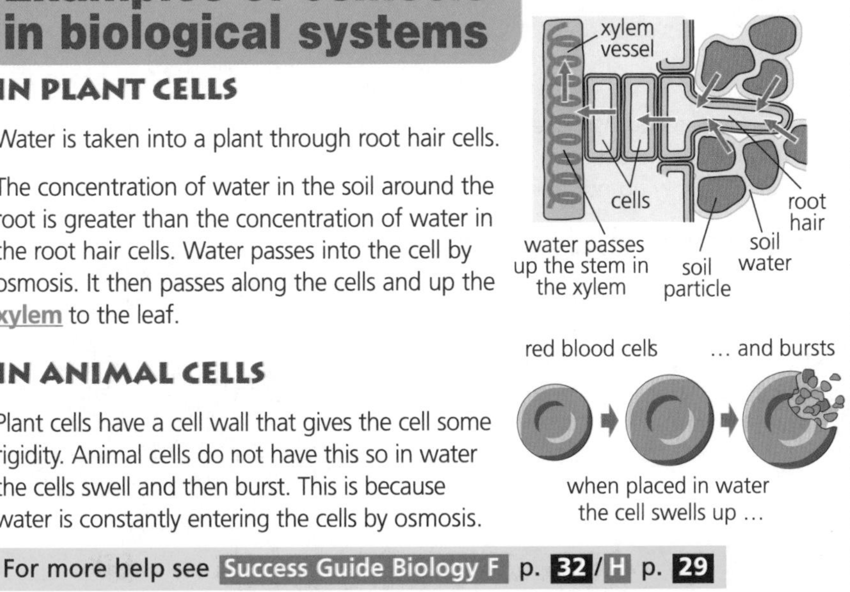

Examples of osmosis in biological systems

IN PLANT CELLS

Water is taken into a plant through root hair cells.

The concentration of water in the soil around the root is greater than the concentration of water in the root hair cells. Water passes into the cell by osmosis. It then passes along the cells and up the xylem to the leaf.

IN ANIMAL CELLS

Plant cells have a cell wall that gives the cell some rigidity. Animal cells do not have this so in water the cells swell and then burst. This is because water is constantly entering the cells by osmosis.

For more help see Success Guide Biology F p. 32 / H p. 29

Rates – concentration

Reaction involving two gases

In diagram 2 there are more particles in a given volume than in diagram 1.

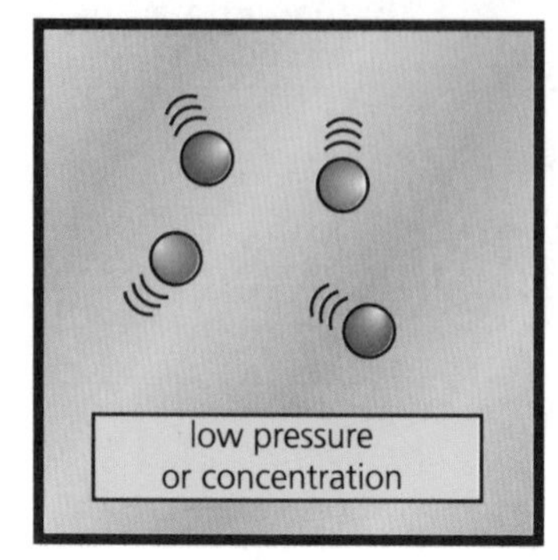

Diagram 1

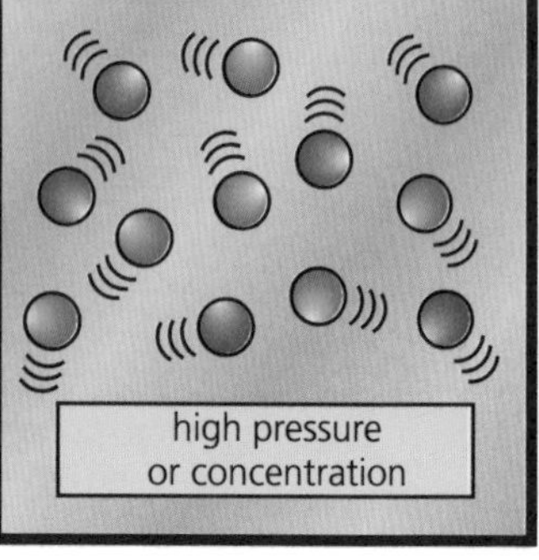

Diagram 2

In diagram 2 the gas has:

- a higher **concentration**
- a higher **pressure**
- been kept at the same temperature as the gas in diagram 1.

The gas is made up of two reactants, A and B.

These two gases react to form C.

$$A + B \rightarrow C$$

The reaction only takes place when particles of A and B collide. However, the colliding particles must have more energy (called the **activation energy**) before they react. If they do not have enough energy they will just bounce apart.

WHY IS THE REACTION FASTER?

The reaction in diagram 2 will be very much faster than in diagram 1 because there is a greater **frequency** of successful collisions (i.e. more collisions per unit of time).

TIP!

Many students write that there are more collisions in diagram 2 than in diagram 1. This is only true if the same time period is considered.

For more help see Success Guide Chemistry F p. 75/H p. 71

Reactions involving solutions

Sodium thiosulphate solution and dilute hydrochloric acid react when mixed.

$$Na_2S_2O_3(aq) + 2\,HCl(aq) \rightarrow S(s) + 2NaCl(aq) + H_2O(l) + SO_2(g)$$

This experiment lasts from when the solutions are mixed until a cross drawn on a piece of paper under the beaker disappears.

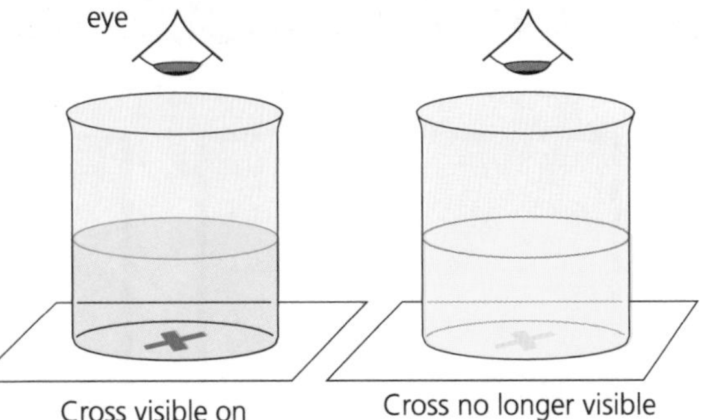

Cross visible on paper through transparent solution

Cross no longer visible as sulphur precipitate makes solution cloudy

When a series of experiments is carried out using the same conditions but only altering the concentration of sodium thiosulphate, the time for the cross to disappear from view decreases as the concentration of sodium thiosulphate increases.

The rate of reaction increases as the concentration increases.

WHY IS THE REACTION FASTER?

Increasing the concentration of sodium thiosulphate increases the number of thiosulphate ions in the solution. There are more successful collisions between the thiosulphate and hydrogen ions from the acid.

The increase in rate is lower with solutions than with gases because the particles do not have as much kinetic energy.

Students using thiosulphate/acid reactions often study the effect of changing the concentration of hydrochloric acid. This is unwise as the chemistry of this is very complex. They would get better results if they studied the effect of changing the concentration of thiosulphate, keeping the concentration of acid constant.

What concentration units should you use?

At GCSE Science the only concentration units you should use are g/dm^3.
In some books you may see concentrations such as 2M. You do not have to use these units.

Rates – temperature

Reaction involving two gases

There are the same number of particles in diagrams 1 and 2.

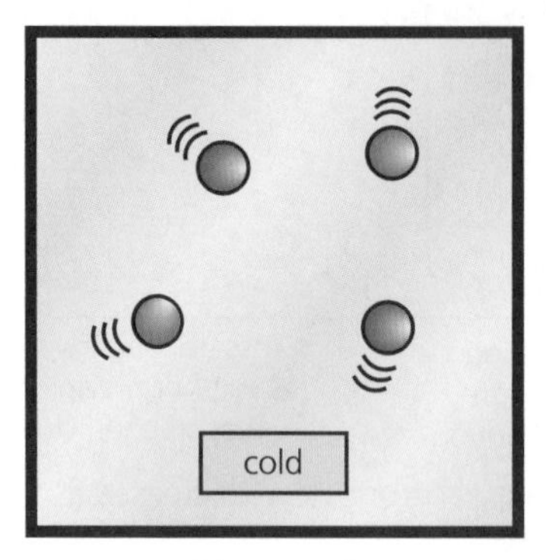

Diagram 1

hot

Diagram 2

The gas is made up of two reactants, A and B.

These two gases react to form C.

$$A + B \rightarrow C$$

The reaction only takes place when particles of A and B collide. However, the colliding particles must have more energy (called the **activation energy**) before they react. If they do not have enough energy they will just bounce apart.

The reaction in diagram 2 is faster than in diagram 1.

WHY IS THE REACTION FASTER?

The reaction in diagram 2 is faster than in diagram 1 because the particles are moving faster, so there is a greater frequency in the collisions between particles of A and B. More collisions will have sufficient energy to exceed the activation energy.

TIP!

Many books state that a 10°C temperature rise doubles the rate of reactions. While this is sometimes the case it is not universally true.

For more help see Success Guide Chemistry F p. 74/H p. 70

Reactions involving solutions

Again, the reaction of sodium thiosulphate solution and dilute hydrochloric acid can be used to investigate the effect of temperature.

$$Na_2S_2O_3(aq) + 2\ HCl(aq) \rightarrow S(s) + 2NaCl(aq) + H_2O(l) + SO_2(g)$$

This time the volumes and concentrations of the two reactants have to be kept the same. The only independent variable is the temperature.

The rate of reaction increases as the temperature increases.

For safety reasons the temperature used should not exceed 55°C. Above this temperature poisonous sulphur dioxide is released.

The reaction of sodium thiosulphate and hydrochloric acid can be studied using a light source above the beaker and a light detector below. The percentage of light transmitted decreases during the reaction and this can be displayed on a computer screen.

WHY IS THE REACTION FASTER?

Increasing the temperature of the solutions increases the frequency of collisions between the thiosulphate and hydrogen ions in the acid. More of these collisions will also have sufficient energy to exceed the activation energy.

Examples where the effect of temperature is important

Milk sours quite quickly when left at room temperature. It keeps much longer if stored in a refrigerator at 5°C. At the lower temperature the reactions are slowed down.

Potatoes cook faster in hot fat at 300°C than in boiling water at 100°C.

In industrial processes, lowering the temperature will reduce energy costs.

Rates – particle size

Explosions in a coal mine

Coal mines are very hazardous places. Lumps of coal do not burn easily in air. However, coal dust mixed with air in a coal mine can explode. This is because coal dust has a very much larger surface area than a lump of coal of the same mass.

Lumps or powder

In diagrams 1 and 2 there are the same amounts of solid and gas. However, the solid in diagram 2 is in much smaller pieces and so has a **larger surface area**.

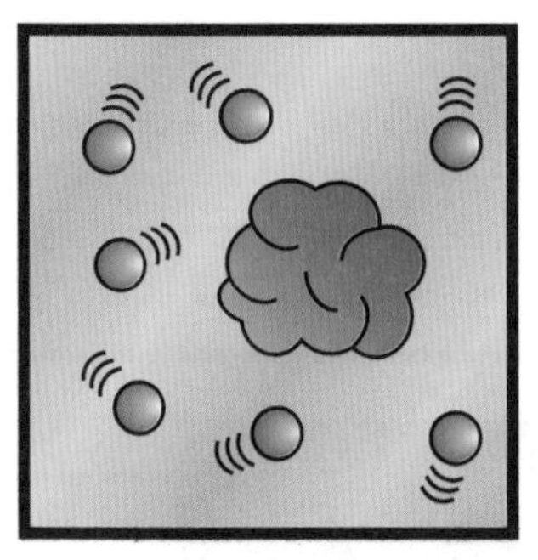

Diagram 1

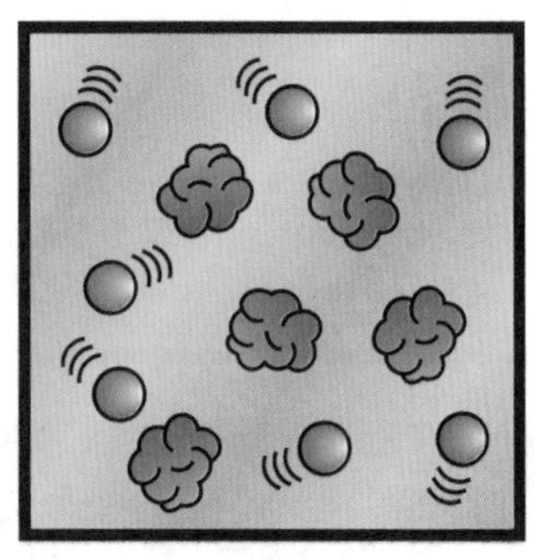

Diagram 2

For the reaction to take place, gas particles must collide with the solid. The same proportion of the collisions will lead to reaction. However, the reaction in diagram 2 is faster. There are more collisions per unit of time because the solid has a larger surface area.

How does surface area change?

If you had a solid cuboid 4 cm × 4 cm × 4 cm it would have a surface area of 64 cm^2. If the same solid is cut into 64 blocks each 1 cm × 1 cm × 1 cm, the total surface area is:

number of blocks x number of faces x area of face = $64 \times 6 \times 1 = 384\ cm^2$.

Imagine what the surface area would be if the block was cut into blocks 0.1 cm × 0.1 cm × 0.1 cm.

For more help see Success Guide Chemistry F p. 74/H p. 70

Calcium carbonate and dilute hydrochloric acid

Marble chips (calcium carbonate) and dilute hydrochloric acid react to produce carbon dioxide gas:

$$CaCO_3(s) + 2HCl(g) \rightarrow CaCl_2(aq) + H_2O(l) + CO_2(g)$$

Equal volumes of hydrochloric acid of the same concentration are used with equal masses of lumps of marble and powder.

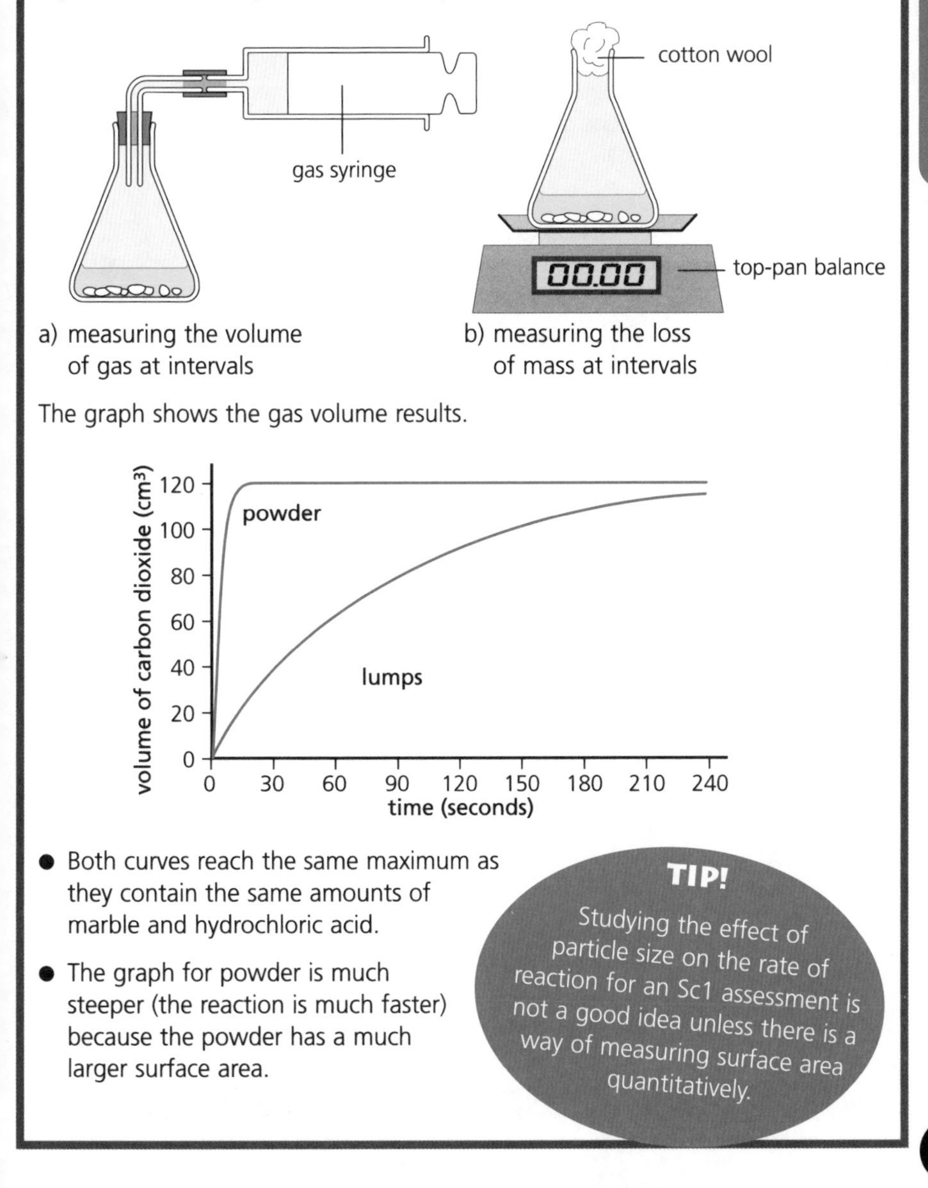

a) measuring the volume of gas at intervals

b) measuring the loss of mass at intervals

The graph shows the gas volume results.

- Both curves reach the same maximum as they contain the same amounts of marble and hydrochloric acid.
- The graph for powder is much steeper (the reaction is much faster) because the powder has a much larger surface area.

TIP!

Studying the effect of particle size on the rate of reaction for an Sc1 assessment is not a good idea unless there is a way of measuring surface area quantitatively.

Rates – catalysts

Catalysts

A catalyst is a substance that increases the rate of a reaction. It is not used up and is specific to a particular reaction.

REMEMBER

A catalyst **speeds up** a reaction. It does not produce more – just the same amount more quickly. Using a catalyst does not produce more product in an industrial process but it does produce the same amount more quickly and therefore more economically.

EXAMPLE

At room temperature, hydrogen peroxide solution decomposes slowly into water and oxygen. With a catalyst the reaction takes place more quickly.

$$2H_2O_2(aq) \rightarrow 2H_2O(l) + O_2(g)$$

One catalyst for this reaction is manganese(IV) oxide, MnO_2.

Although the mass of catalyst is unchanged it may change from a solid lump to a powder.

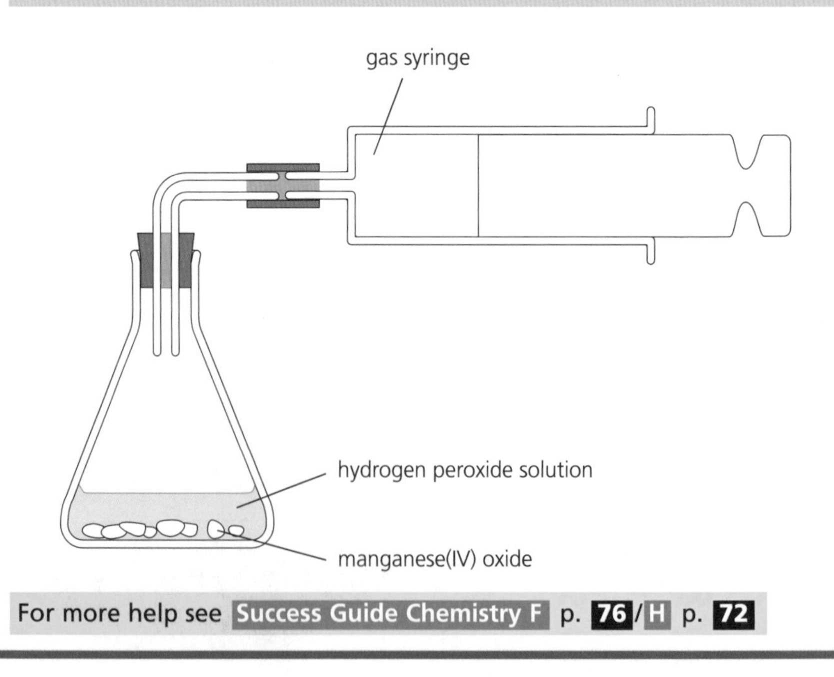

For more help see Success Guide Chemistry F p. 76 / H p. 72

Energy level diagram

The diagram shows an energy level diagram for an **exothermic reaction**.

Notice that the energy of the products is less than the energy of the reactants.

The energy given out, ΔH, is the difference between these two levels.

The red graph shows the progress of the reaction without a catalyst. Before reaction takes place the collisions between reactant particles have to have enough energy to get over the **activation energy barrier**.

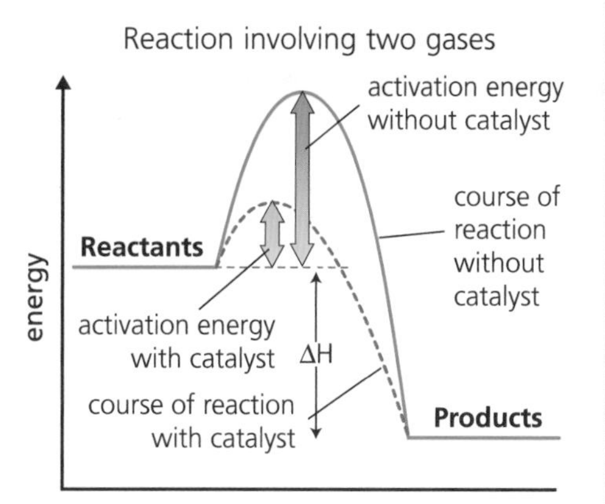

The blue graph shows the progress of the reaction when a catalyst is used. Now, the reaction has a lower activation energy. More collisions of reactant particles get over this lower activation reaction. As a result the reaction is faster.

Transition metals as catalysts

There are many examples of reactions that are catalysed. In most cases, the catalyst is either a transition metal or a compound of a transition metal.

In the Haber process, nitrogen and hydrogen combine to form ammonia. $N_2(g) + 3H_2(g) \rightleftharpoons 2NH_3(g)$. The catalyst for this reaction is iron.

In a **catalytic converter** in a car exhaust system, nitrogen oxides and carbon monoxide are converted to nitrogen and carbon dioxide. The catalyst is platinum.

In the contact process, sulphuric acid is produced using a vanadium(V) oxide catalyst.

Catalysts do not last for ever

Although catalysts are not used up they do not last indefinitely in an industrial process. They have to be replaced from time to time.

The catalyst can be poisoned. Using lead in petrol will quickly poison the platinum catalyst in a catalytic converter and stop it working.

At high temperatures catalysts can sometimes be broken down physically.

For more help see **Success Guide Chemistry F** p. **76**/**H** p. **72**

Rates – enzymes

Enzymes

An enzyme is a biological catalyst. It is active only under certain conditions of temperature and pH specific to a particular reaction. It is a protein.

EXAMPLE

Fermentation (anaerobic respiration) of glucose solution to produce ethanol and carbon dioxide is catalysed by the enzyme zymase in yeast.

$$C_6H_{12}O_6(aq) \rightarrow 2C_2H_5OH\ (aq) + 2CO_2(g) + \text{energy}$$

Fermentation takes place in warm conditions and stops when about 14% ethanol is produced. At this stage, the ethanol poisons the enzyme.

Temperature

Most enzymes work best at temperatures between 30–40°C.

If the temperature is too low the enzymes are inactive.

At temperatures above about 40°C the enzyme is permanently damaged or **denatured**.

This is a typical graph for the activity of an enzyme at different temperatures. At the peak the enzyme is more effective.

enzyme activity

0 20 40 60 80

temperature (°C)

Enzyme washing powders are not effective in cold water or very high temperature washes.

TIP!

Many students write that an enzyme is a living catalyst and is killed at a high temperature. This is not credit worthy.

TIP!

If attempting an investigation of the effect of temperature on the action of an enzyme, even five sets of results will not give you confidence in any curve you draw.

For more help see Success Guide Chemistry F p. 76–7 / H p. 72–3

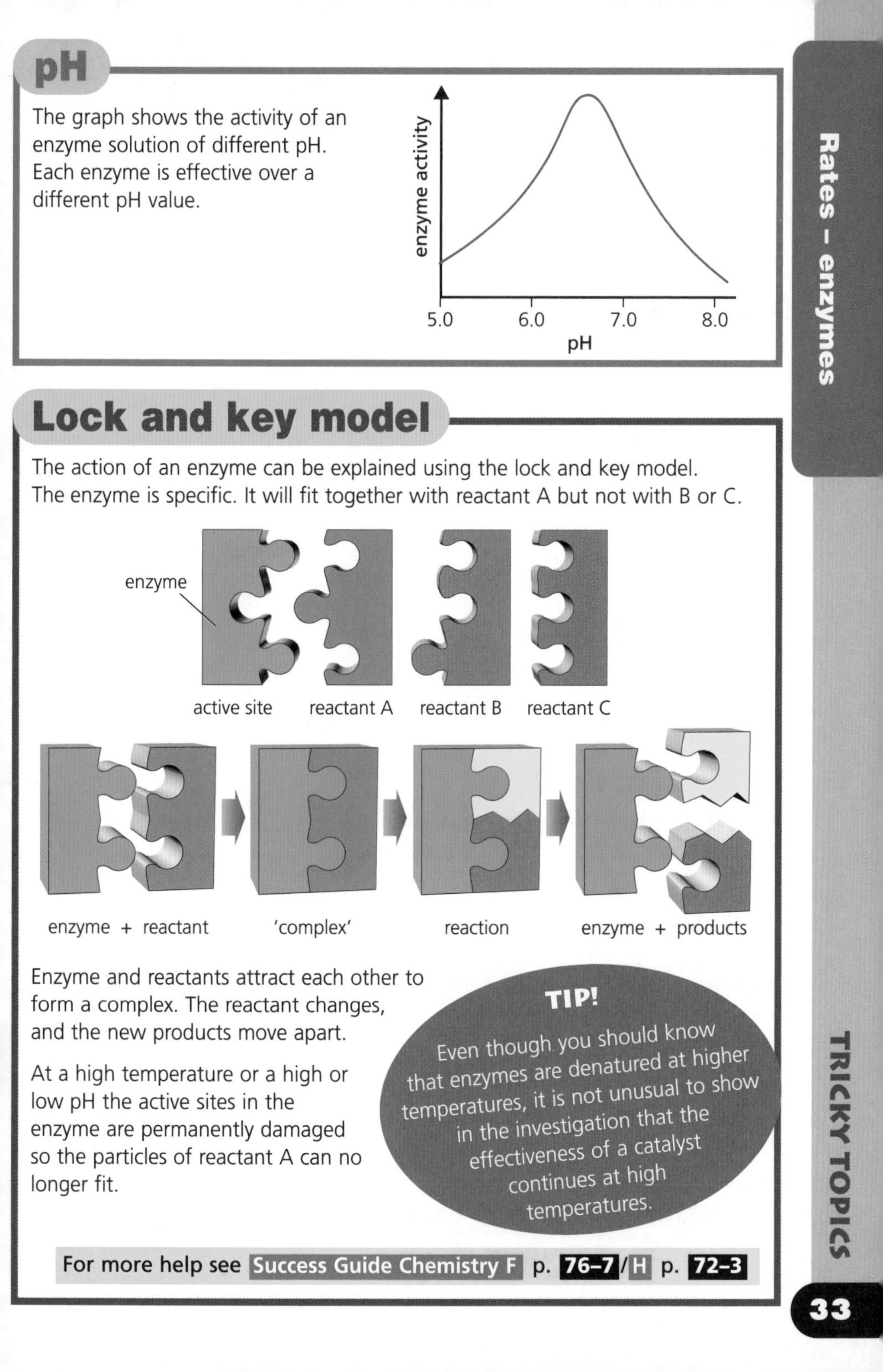

pH

The graph shows the activity of an enzyme solution of different pH. Each enzyme is effective over a different pH value.

enzyme activity
5.0
6.0
7.0
8.0
pH

Lock and key model

The action of an enzyme can be explained using the lock and key model. The enzyme is specific. It will fit together with reactant A but not with B or C.

enzyme
active site
reactant A
reactant B
reactant C

enzyme + reactant
'complex'
reaction
enzyme + products

Enzyme and reactants attract each other to form a complex. The reactant changes, and the new products move apart.

At a high temperature or a high or low pH the active sites in the enzyme are permanently damaged so the particles of reactant A can no longer fit.

TIP!

Even though you should know that enzymes are denatured at higher temperatures, it is not unusual to show in the investigation that the effectiveness of a catalyst continues at high temperatures.

For more help see **Success Guide Chemistry F** p. **76–7**/**H** p. **72–3**

Atomic structure

You will need to understand atomic structure both for physics and chemistry. In physics the emphasis is on the nucleus and in chemistry on the electrons that move around the nucleus, especially outer electrons.

Particles in atoms

Two hundred years ago scientists like John Dalton believed that atoms were indivisible.

We now know that atoms of all elements are made up of three particles – **protons**, **electrons** and **neutrons**.

The properties of the particles are summarised in the table.

particle	mass	charge
proton p	1 amu	+1
neutron n	1 amu	0
electron e	negligible	–1

As all atoms are **neutral**, each atom must contain an equal number of **protons and electrons**.

The protons and neutrons, together called **nucleons**, are packed together in the positively changed **nucleus**.

The electrons move around the nucleus in clearly defined **shells** or **energy levels**.

Mass number and atomic number

A lithium atom can be shown as:

mass or nucleon number → 7
atomic or proton number → 3

$^{7}_{3}Li$

Using the information, we can work out that a lithium atom contains 3 protons, 3 electrons and 4 neutrons.

For more help see Success Guide Physics F p. 81 / H p. 78

For more help see Success Guide Chemistry F/H p. 48

Electron arrangement

Here is a simple model of an atom showing shells of electrons. The shells fill up in order of increasing energy, the ones closest to the nucleus first.

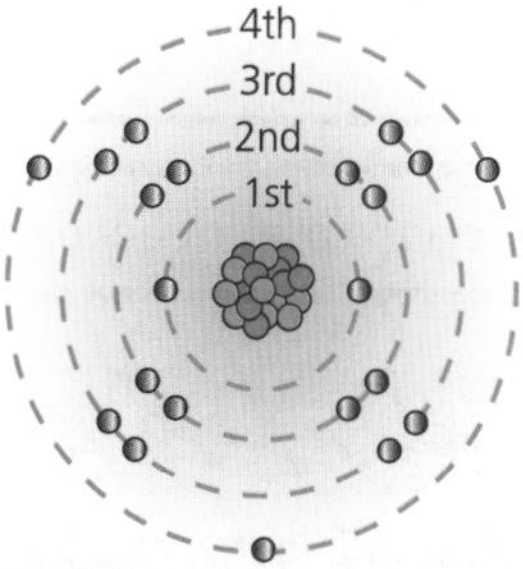

4th
3rd
2nd
1st

The relationship between electron arrangement and the Periodic Table

The diagrams show simple diagrams for lithium, magnesium and chlorine atoms.

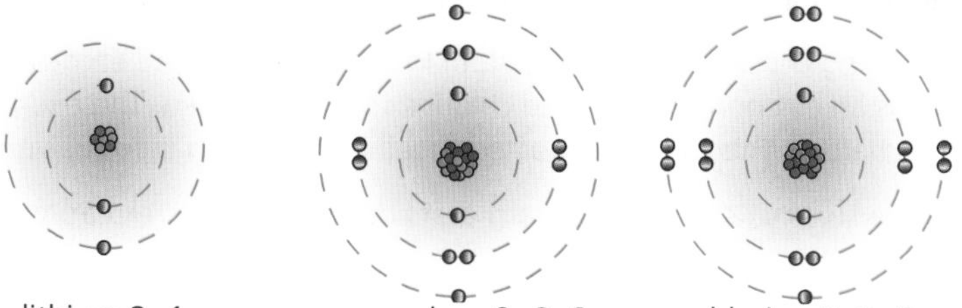

	lithium	magnesium	chlorine
electron arrangement	2, 1	2, 8, 2	2, 8, 7
group in Periodic Table	1	2	7

The number of electrons in the outer shell of an atom is the same as the group number in which the element is placed.

Ions

Ions are formed when **metal atoms lose** outer electrons or **non-metal atoms gain** electrons.

The metal ions are positively charged and the non-metal ions are negatively charged. Lithium ion Li^+ contains 3 protons and 2 electrons – one fewer than a lithium atom. Chloride ion Cl^- contains 17 protons and 18 electrons – one more than a chlorine atom.

The electron arrangement of the first 20 elements is listed on page 10 and a copy of the Periodic Table is on page 8.

For more help see Success Guide Physics F p. 83/H p. 79

For more help see Success Guide Chemistry F p. 50/H p. 49

Bonding

Ionic and covalent bonding are two types of bonding.

Ionic bonding

Ionic bonding involves the **complete transfer** of one or more electrons from a metal to a non-metal atom.

This happens in such a way that the ions formed have completely **full outer shells**.

SODIUM CHLORIDE

The electron arrangements in sodium and chlorine atoms are 2, 8, 1 and 2, 8, 7.

When bonding occurs, each sodium atom loses an electron and each chlorine atom gains an electron. Sodium and chloride **ions** are formed. These ions have opposite charges. The ions are attracted together by **strong electrostatic forces**.

This can be summarised by a diagram that shows only outer electrons.

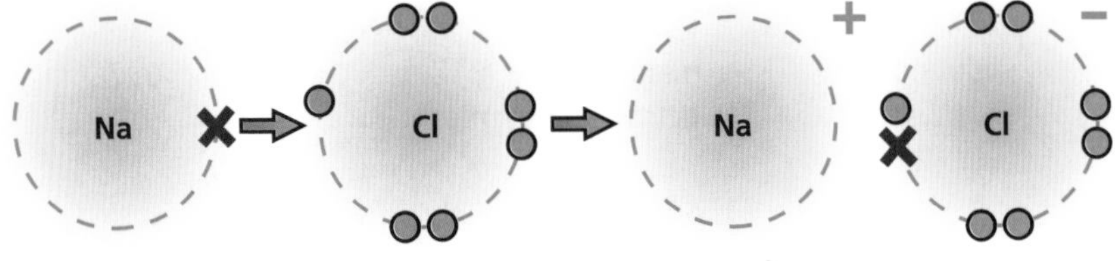

MAGNESIUM OXIDE

When magnesium and oxygen combine, each magnesium atom loses two electrons and each oxygen atom gains two electrons.

The diagram summarises this change.

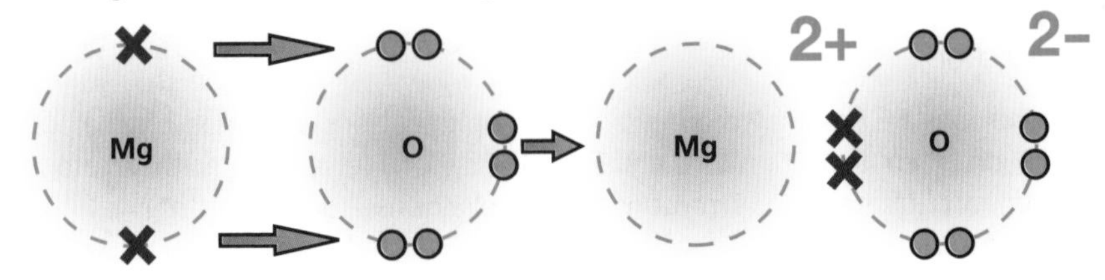

Ionic compounds exist as lattices of regularly arranged ions.

Magnesium oxide has a much high melting point than sodium chloride because the charges are 2+ and 2– instead of 1+ and 1–.

For more help see Success Guide Chemistry F p. 52–3/H p. 50

Covalent bonding

Covalent bonding involves a **sharing** of pairs of electrons in bonds between non-metal atoms.

Covalent bonding allows both atoms in a bond to have a stable, full outer shell.

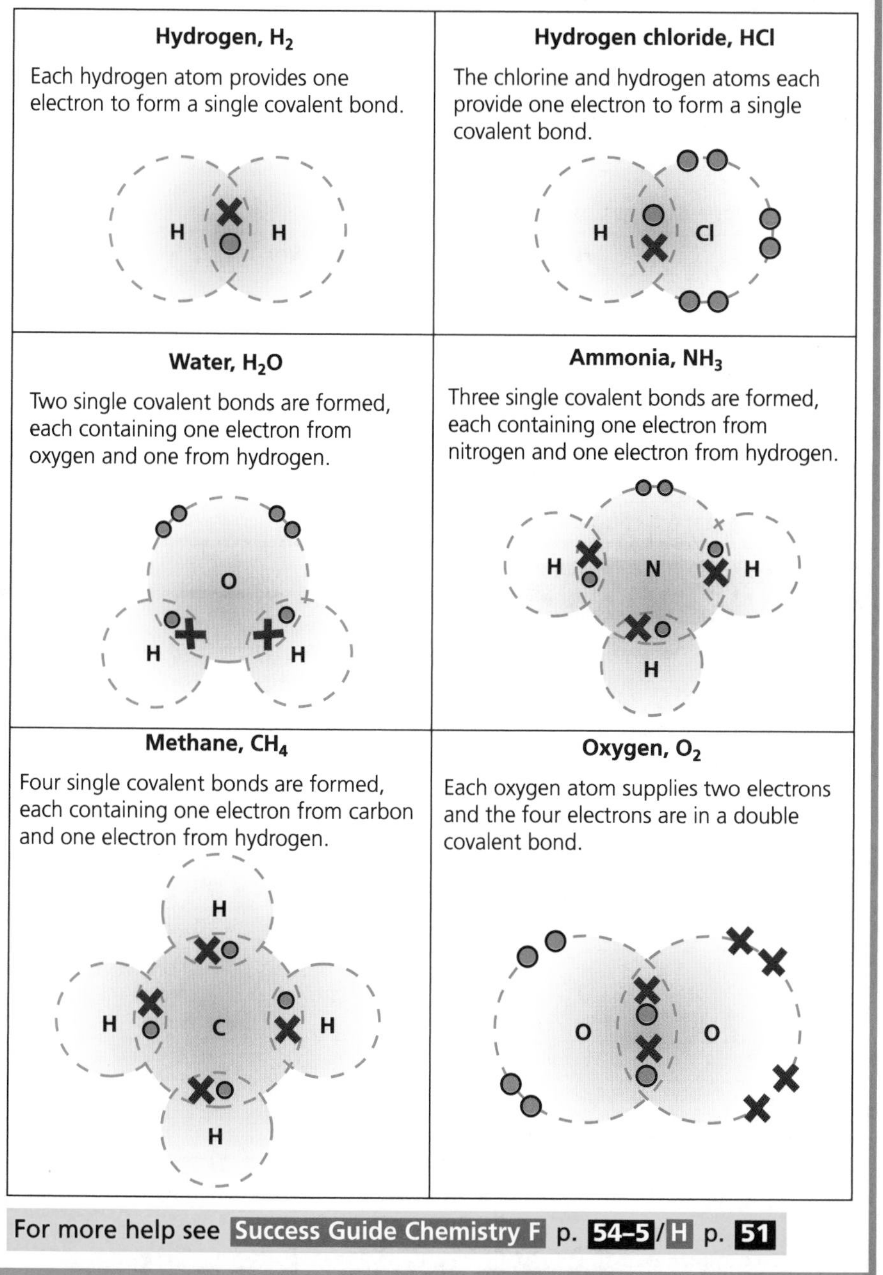

Hydrogen, H_2

Each hydrogen atom provides one electron to form a single covalent bond.

Hydrogen chloride, HCl

The chlorine and hydrogen atoms each provide one electron to form a single covalent bond.

Water, H_2O

Two single covalent bonds are formed, each containing one electron from oxygen and one from hydrogen.

Ammonia, NH_3

Three single covalent bonds are formed, each containing one electron from nitrogen and one electron from hydrogen.

Methane, CH_4

Four single covalent bonds are formed, each containing one electron from carbon and one electron from hydrogen.

Oxygen, O_2

Each oxygen atom supplies two electrons and the four electrons are in a double covalent bond.

For more help see Success Guide Chemistry F p. 54–5/H p. 51

Radioactivity

Isotopes

Most elements are made up from more than one type of atom. Atoms of the same element containing the same number of protons and electrons but different numbers of neutrons are called **isotopes**.

Isotopes have different physical properties but similar chemical properties.

There are three isotopes of hydrogen.

name	Hydrogen	Deuterium	Tritium
symbol	$^{1}_{1}H$	$^{2}_{1}H$	$^{3}_{1}H$
protons	1	1	1
electrons	1	1	1
neutrons	0	1	2
radioactive	no	no	yes

Some elements, e.g. fluorine, exist as a single isotope.

Radioactivity

Radioactivity is the spontaneous breakdown of the nucleus of atoms leading to the emission of radioactive particles or electromagnetic radiation.

Most radioactive isotopes have nuclei containing large numbers of protons and neutrons. These are more likely to be unstable and break down.

There are three types of emission from radioactive materials: alpha (α), beta (β) and gamma (γ).

The table compares these three types of radiation.

radiation	alpha	beta	gamma
nature	slow-moving helium nuclei, i.e. 2p and 2n	fast-moving electrons	short wavelength electro-magnetic waves
mass	4	negligible	0
charge	+2	–1	0
relative ionising power	100 000	1 000	1
penetrating power in air	1–5 cm	10–80 cm	almost unlimited

For more help see Success Guide Physics F p. 83–5/H p. 79–81

Deflection and penetration

The diagrams show the deflection of each type of radiation in a magnetic field (right) and the penetration through different materials (below).

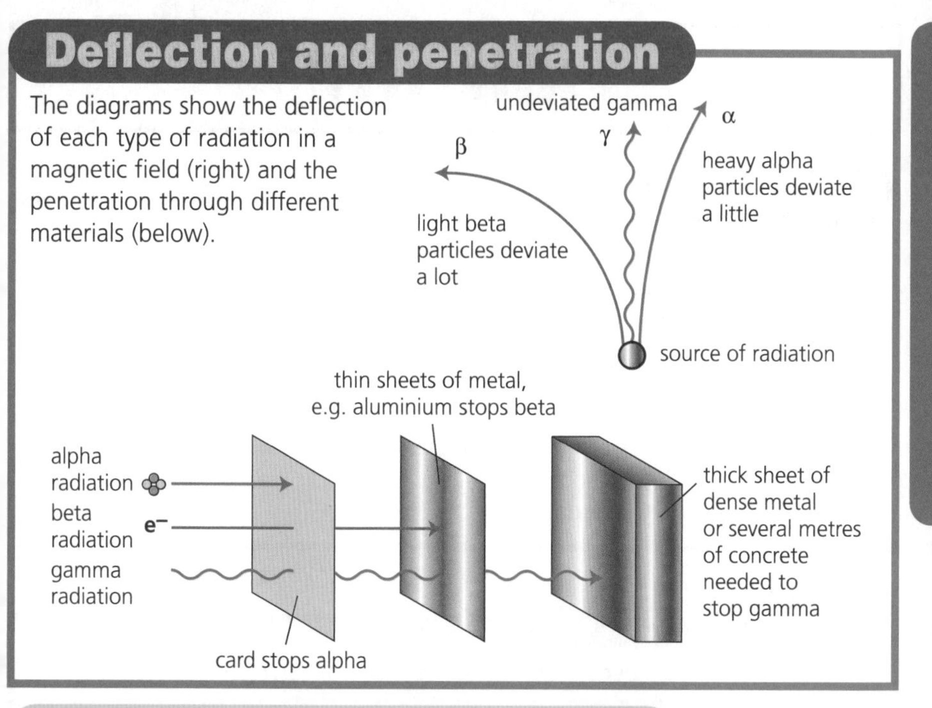

Changes that take place in radioactive decay

ALPHA EMISSION

e.g. $${}^{226}_{88}Ra \rightarrow {}^{222}_{86}Rn + {}^{4}_{2}He$$

Radon (Rn) is two places to the left of Radium (Ra) in the Periodic Table.

BETA EMISSION

e.g. $${}^{14}_{6}C \rightarrow {}^{14}_{7}N + {}^{0}_{-1}e$$

A neutron in the nucleus changes to a proton and an electron. This electron is lost and is not an outer electron. Nitrogen is one place to the right of carbon in the Periodic Table.

GAMMA EMISSION

When an unstable nucleus emits alpha or beta particles, excess energy is emitted as gamma radiation.

For more help see Success Guide Physics F p. 84–5/H p. 81–3

Half-life

Half-life

The activity of a radioactive source (i.e. the amount of radiation emitted per unit time) depends upon the number of unstable nuclei in the sample. Activity therefore decreases with time.

The half-life is the time taken for the number of undecayed nuclei in a sample of a radioactive isotope to halve.

The half-lives of isotopes can be very different, e.g.

Uranium-238 4500 million years
Radium-214 20 minutes.

Half-life curves

All radioactive sources decay with the same shape of curve. This can be used to find the half-life of the isotope.

The graph shows a typical half-life curve.

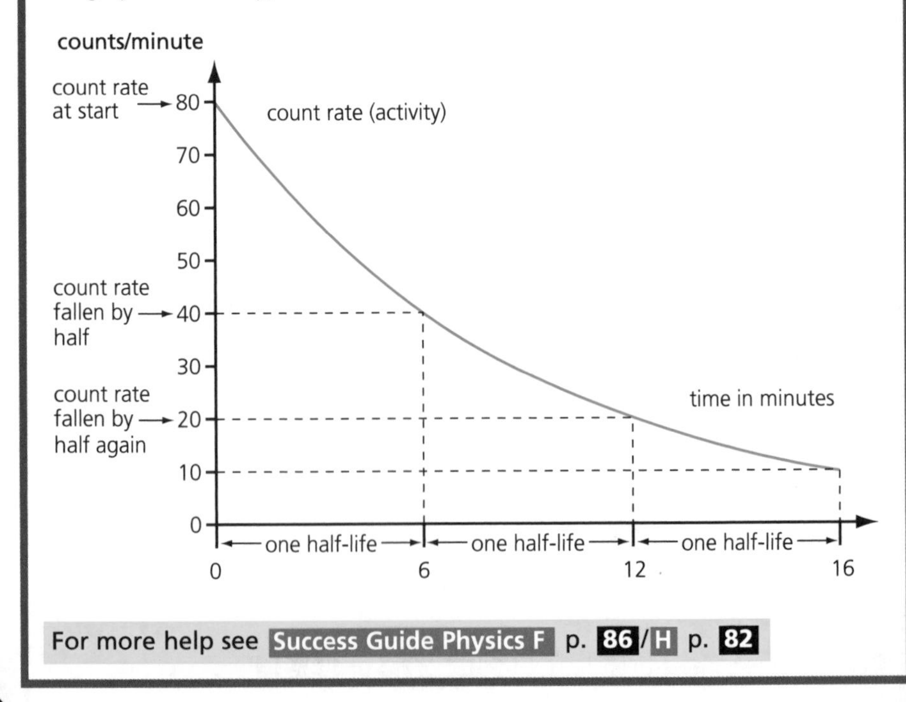

For more help see Success Guide Physics F p. 86/H p. 82

Count rate

The time taken for the count rate (measured using a Geiger counter) to drop from 80 counts/minute to 40 counts/minute (to halve) is 6 minutes. This is the half-life. You will get the same result if you start at another point, e.g. 40 to 20.

TIPS!

Don't fall into the trap of thinking if the half-life is 6 minutes all of the sample will have decayed after 12 minutes. After 12 minutes a quarter will remain and after 18 minutes an eighth will remain. It takes a very long time for all to decay.

When working out the half-life of a radioactive isotope, choose a point high up on the curve and show the dotted lines. They help the examiner to work out what you are doing.

Fission or fusion?

The breaking down of an isotope by radioactive decay is called **nuclear fission**. Fission occurs in a nuclear reactor to release energy, e.g.

$$^{235}_{92}\text{U} + ^{1}_{0}\text{n} \rightarrow \text{X} + \text{Y} + 2 \text{ or } 3 \text{ neutrons} + \text{energy}$$

(X + Y: 2 smaller nuclei)

It is possible to set up a **chain reaction**.

The opposite of nuclear fission is **nuclear fusion**. Small nuclei join together to form a heavier, more stable nucleus. This process releases a large amount of energy and usually requires a large amount of energy to start it off.

Nuclear fusion reactions take place within the Sun and cause the temperature of the Sun to be several million °C.

A typical fusion reaction is shown below.

Some scientists have claimed that nuclear fusion reactions have been brought about at room temperature. However, the experiments could not be replicated.

For more help see Success Guide Physics F p. 86/H p. 86–7

Life of a star

Our Sun is a star and a small relatively young star – about five thousand million years old.

Stars are formed in enormous clouds of dust and hydrogen gas when the attractive force between particles in the clouds cause them to collapse. Nuclear fusion reactions begin (see page 41) creating helium and releasing large amounts of energy.

The star then enters its main sequence.

Life of a small star

The diagram shows the life cycle of a small star such as our Sun.

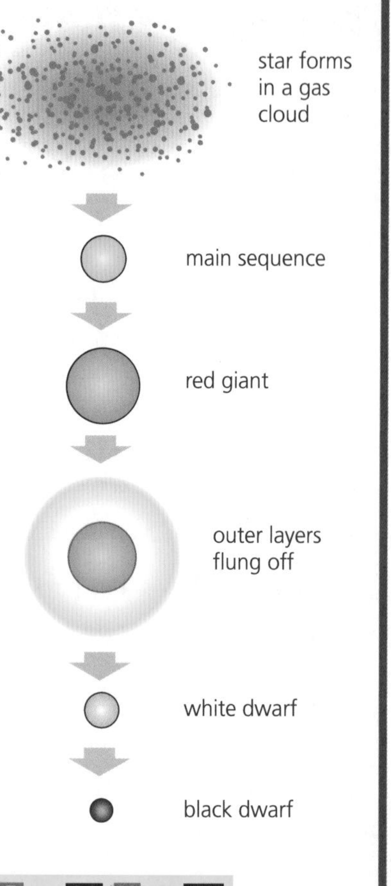

Our Sun is currently in its **main sequence**.

Energy is released by the Sun through the fusion of hydrogen nuclei.

Outward forces caused by the high pressure in the core are balanced by gravitational forces.

When all the hydrogen has been used up the Sun will cool and expand and become a **red giant**.

As the Sun expands the core will contract. The temperature will then become hot enough for fusion of helium nuclei, forming nuclei of carbon and oxygen.

The Sun will then contract, losing its outer layers and becoming a very hot, dense body called a **white dwarf**.

As energy is no longer being generated, the colour of a white dwarf changes as it cools. It eventually becomes an invisible **black dwarf**.

For more help see Success Guide Physics F p. 91/H p. 90

Life of a massive star

The diagram shows the life cycle of a massive star.

More massive stars, after the main sequence, expand to become red **supergiants**.

In a red **supergiant**:

Nuclear fusion in the contracting core results in the formation of nuclei of elements such as magnesium, silicon and iron.

The star then generates energy again and becomes a **blue supergiant**.

When the nuclear reactions are finished the star cools and contracts again, glowing brightly as its temperature increases. It explodes and becomes a **supernova**.

The supernova explodes, flinging off the outer layers to form a **dust cloud**.

The core that is left behind is a **neutron star** or **black hole**.

Very dense neutron stars are called black holes because they are so dense that not even light can escape from their gravitational fields. Black holes are detected by their effect on surrounding objects.

The existence of heavy elements such as iron in inner planets is evidence that our Solar system was formed from the gas and dust flung off from the outer layers of supernovas.

star forms in a gas cloud

main sequence

red supergiant

blue supergiant

exploding supernova

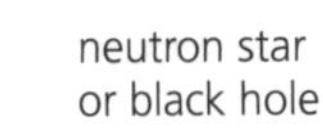

For more help see Success Guide Physics F p. 91/H p. 90

Metals as conductors

Metals are good conductors.
They conduct heat and they conduct electricity.

Structure of metals

The conductivity of metals can be explained by considering the structure of metals.

The diagram shows a metal structure. It is based upon **close packing**.

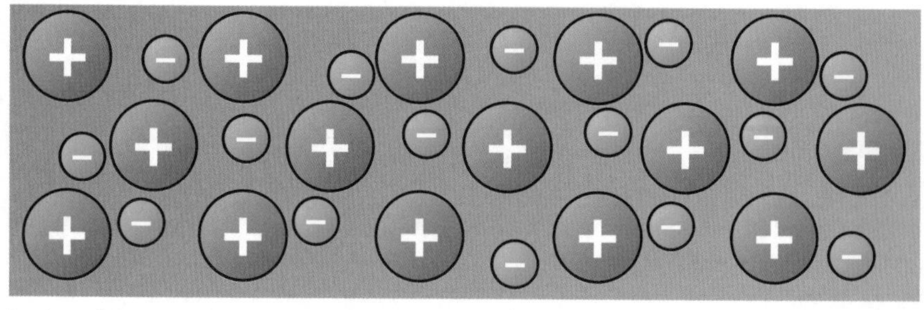

In the **giant structure**, **metal ions** are closely packed together. These positive ions are held together by **electrons** that are free to move throughout the structure. These electrons are **delocalised**.

Explaining the physical properties of metals

The table shows some of the physical properties of metals and the explanation for these properties.

Physical property	Explanation
high density	Close packing means more particles in a given volume
high melting and boiling points	Strong forces hold the structure together
conduct electricity	Delocalised electrons are able to move through the metal structure and carry charge
conduct heat	In hot metal, particles vibrate more. These vibrations pass along from particle to particle
ductile (drawn into wires) or **malleable** (beaten into sheets)	The particles in the structure are able to slide over each other

Alkali metals have a much lower density than other metals. In fact most will float on water. This is because in alkali metals the ions are not closely packed.

For more help see Success Guide Chemistry F/H p. 30

Graphite and diamond

Graphite and diamond are two forms of the element carbon.

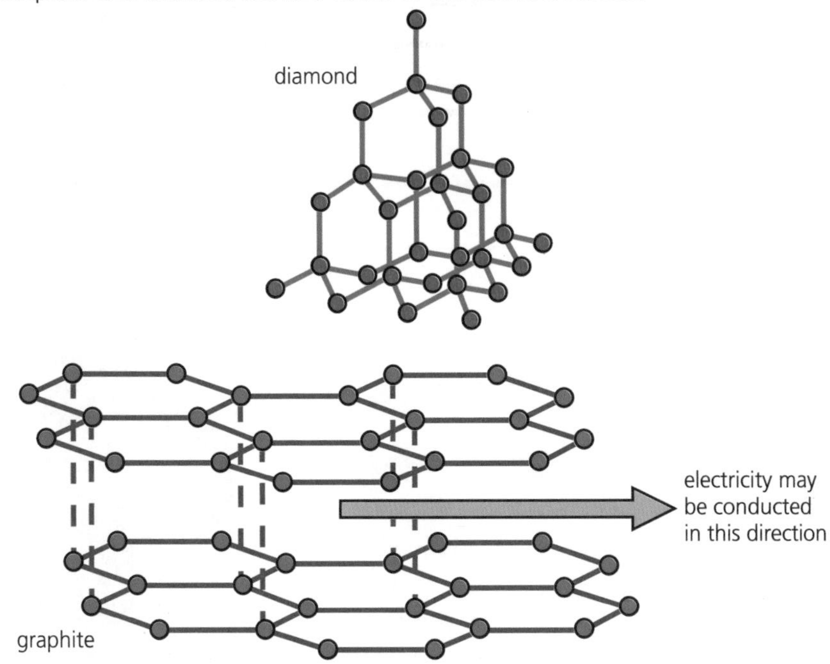

In both forms the carbon atoms are joined together by covalent carbon-carbon bonds.

Graphite is a good conductor of electricity and diamond is a poor conductor of electricity (**insulator**).

The difference can be explained by the structure of these forms of carbon.

In diamond all of the outer electrons are used in covalent bonds between the carbon atoms. These electrons are fixed and cannot move.

In graphite each carbon has an electron that is not used in a covalent bond and these free or delocalised electrons can carry an electric charge through the structure.

Polyethyne

In 2000, two chemists, Hideki Shirakawa and Alan MacDiarmid, were awarded the Nobel Prize for their discovery of a hydrocarbon addition polymer that conducts electricity. This polymer (called polyethyne) and similar materials, may replace metals in electrical circuits.

For more help see Success Guide Chemistry F p. 15/H p. 14

Series and parallel

Electrical components can be connected together in **series** or in **parallel**.

Series

The diagram shows two lamps connected in series to a battery, together with an ammeter that measures current and a voltmeter that measures voltage across one of the lamps.

TIP!

In the series circuit the ammeter is in series but the voltmeter is in parallel with the lamp.

In a series circuit:

- the current is the same at all points in the circuit
- the sum of the voltages across the different components is equal to the voltage of the power supply.

Parallel

The diagram shows two lamps connected in parallel to a battery with an ammeter that measures the total current in both lamps.

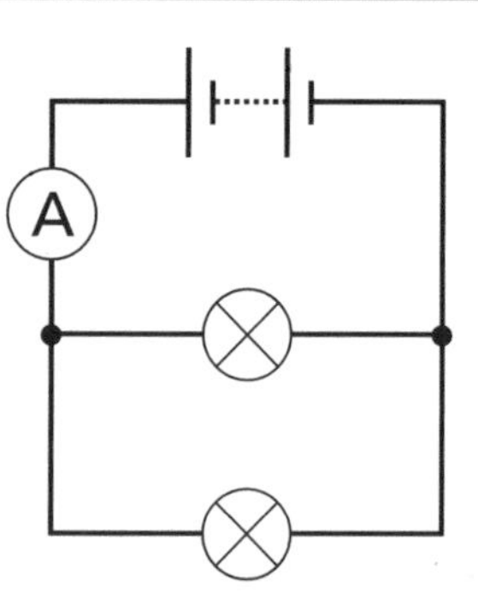

In a parallel circuit:

- all components in parallel have the **same voltage** across them
- the current splits and rejoins at the junctions
- the total current passing into each junction is equal to the current passing out of the junction.

For more help see Success Guide Physics F p. 60–1/H p. 58–9

Resistors in series and parallel

The diagram shows three resistors R_1, R_2 and R_3 connected first in series and then in parallel.

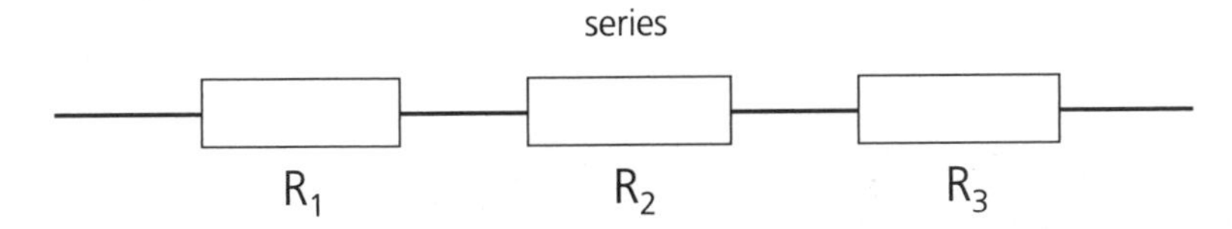

The total resistance, R_S, when the three resistors are in series is:

$$R_S = R_1 + R_2 + R_3$$

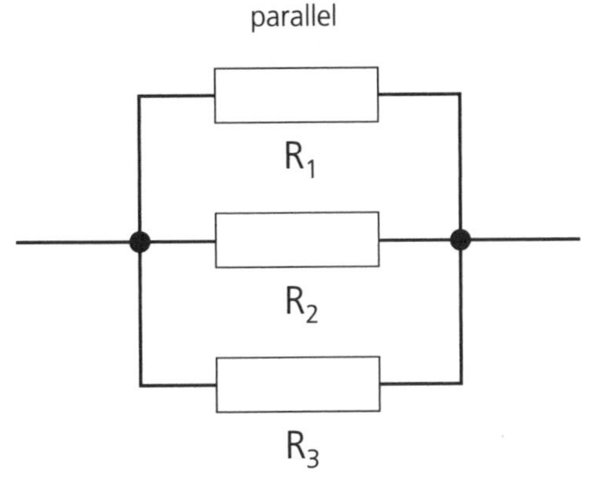

The total resistance, R_p, when the three resistors are in parallel is:

$$\frac{1}{R_P} = \frac{1}{R_1} + \frac{1}{R_2} + \frac{1}{R_3}$$

E.g. if the three resistors are 2Ω , 4Ω and 8Ω:

$$R_s = 2 + 4 + 8 = 14\Omega$$

$$\frac{1}{R_P} = \frac{1}{2} + \frac{1}{4} + \frac{1}{8} = 0.875$$

$$R_P = 1.1\Omega$$

TIP!

The effective resistance of a number of resistors in parallel is always less than the resistance of the smallest resistor.

For more help see Success Guide Physics F p. 62 / H p. 60

Resistance

The current in a circuit depends on the resistance and the voltage.

Measuring resistance

The diagram shows an electrical circuit that can be used to find the resistance of a piece of wire.

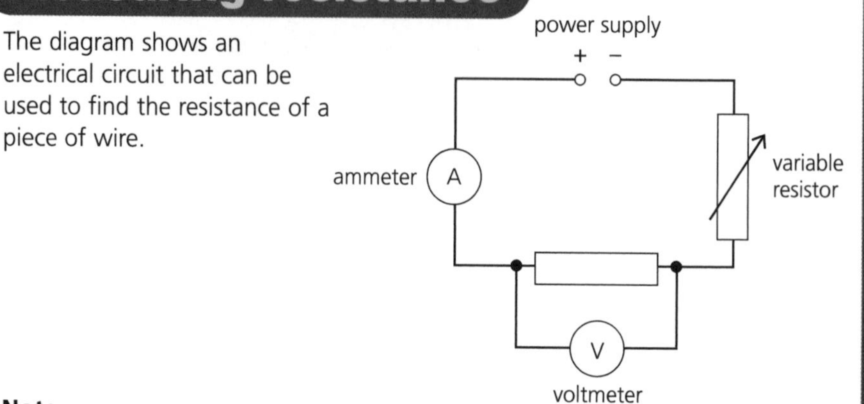

Note

1. The supply voltage is fixed but the current in the circuit can be varied by altering the **variable resistor**.
2. The ammeter is in series but the voltmeter is in parallel.

The graphs show the relationships between voltage and current and resistance and current.

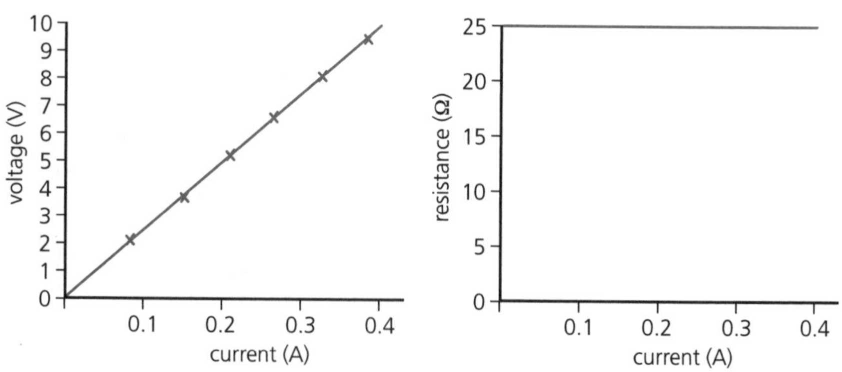

The left-hand graph shows that voltage is directly proportional to current.
The right-hand graph shows that the resistance is constant as current changes.

For more help see Success Guide Physics F p. 62/H p. 60

Ohm's law

These two relationships can be summarised by **Ohm's law**.

$\frac{V}{I} = R$ where V is voltage in **volts (V)**
I is current in **amps (A)**
R is resistance in **ohms (Ω)**

Some students use a triangle to help them remember this relationship.

However, if you draw these triangles in an examination it is not enough to score a mark for the relationship. You must be able to write:

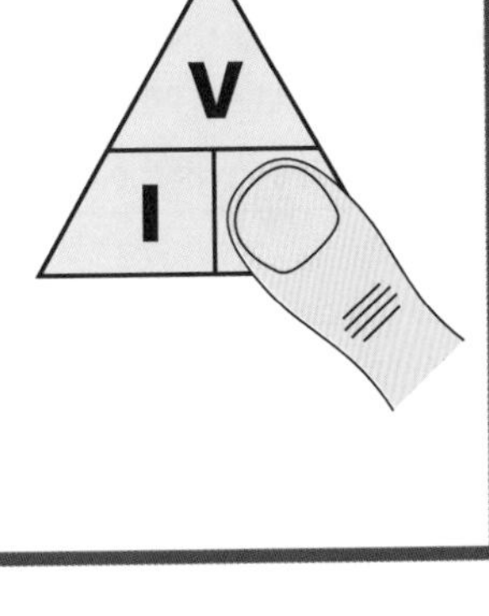

$$\frac{V}{I} = R, V = IR, I = \frac{V}{R}$$

Sample calculation

A car headlamp bulb has a voltage of 12V and a resistance of 2.4Ω. Calculate the current passing.

$I = \frac{V}{R}$ $I = \frac{12}{2.4}$ = **5A**

The resistance of a piece of wire depends on:

- the length of wire – the longer the wire the greater the resistance
- the cross-sectional area of the wire – the larger the cross-sectional area the smaller the resistance.

These can be explained using the structure of a metal (page 44).

WHERE THE EXPERIMENT CAN GO WRONG

Ohm's law is true provided that the temperature of the wire does not change. The graphs show what happens when the filament wire in a light bulb is heated by a passing current.

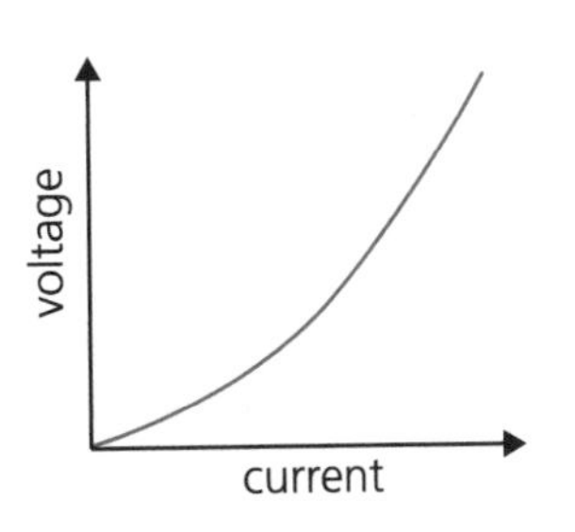

The increasing gradient of the voltage-current graph shows that the resistance of wire increases as the current increases.

For more help see Success Guide Physics F p. 63/H p. 61

The composition of oceans

Two thirds of the Earth's surface is covered by oceans. **The composition of all the oceans is fairly constant**.

The table shows the concentrations of the common ions in the oceans.

Ion	Concentration In g/1000 g of sea water
Chloride Cl^-	19.7
Sodium Na^+	10.7
Sulphate SO_4^{2-}	2.7
Magnesium Mg^{2+}	1.4
Calcium Ca^{2+}	0.4
Potassium K^+	0.4
Hydrogencarbonate HCO_3^-	0.1
Bromide Br^-	0.07

How do these ions get into sea water?

Rain water filters through rocks in the Earth and **dissolves** soluble or sparingly soluble rocks such as halite (sodium chloride) and gypsum (calcium sulphate). Other rocks, such as limestone, do not dissolve in rain water but **react** with water in the presence of carbon dioxide to form carbonic acid.

$$CaCO_3(s) + H_2O(l) + CO_2(g) \rightleftharpoons Ca(HCO_3)_2(aq)$$

In the rivers and oceans, water evaporates but the soluble ions remain.

WHY DOES THE CONCENTRATION OF IONS REMAIN CONSTANT?

The concentration of ions in sea water might be expected to increase as water evaporates and more ions enter as rivers empty water into the oceans.

There must be processes that remove ions from the sea and keep the concentration approximately constant.

Concentration of ions

1 PRECIPITATION

The concentration of calcium ions might be expected to be higher as calcium rocks are very common in the Earth's crust.

However, when the concentrations of certain ions reach a certain level, **precipitation** occurs.

Calcium ions and sulphate ions react to form **calcium sulphate**, which is only sparingly soluble.

$$Ca^{2+}(aq) + SO_4^{2-}(aq) \rightarrow CaSO_4(s)$$

2 FORMING SEA SHELLS

Many marine animals have calcium carbonate in their shells. As these animals develop they have to take in calcium and carbonate ions to make calcium carbonate.

$$Ca^{2+}(aq) + CO_3^{2-}(aq) \rightarrow CaCO_3(s)$$

This process reduces the amount of calcium and carbonate in water.

TIP!

Carbonate comes from the decomposition of calcium hydrogencarbonate.

Extracting salt from sea water.

THE OCEANS ARE A SOURCE OF CHEMICALS

In various parts of the world the oceans are used to extract minerals. Evaporation of sea water produces solid sea salt. Magnesium and bromine are extracted in appreciable amounts.

Electromagnetic radiation

There is a family of electromagnetic waves that is sometimes called the electromagnetic spectrum.

This is shown in the diagram.

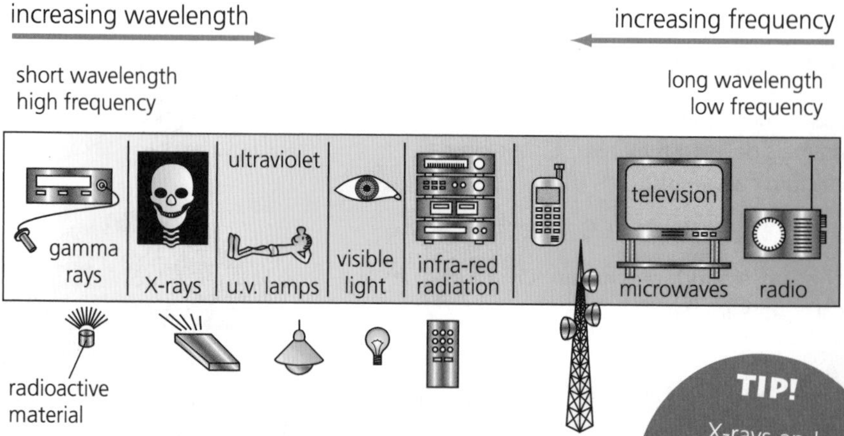

increasing wavelength →

← increasing frequency

short wavelength
high frequency

long wavelength
low frequency

gamma rays | X-rays | ultraviolet u.v. lamps | visible light | infra-red radiation | television microwaves | radio

radioactive material

TIP! X-rays and gamma rays are ionising radiations (see pages 38–9).

All electromagnetic waves have some things in common.

- They are all **transverse waves**.
- They all **transfer energy**.
- They can **travel through a vacuum**.
- They all **travel at the same speed** through a vacuum (300 000 000 m/s)
- They can be **reflected**, **refracted** and **diffracted**.

However, they have different uses because of different **wavelengths** and **frequencies**.

The diagram shows the visible part of the electromagnetic spectrum.

	red	orange	yellow	green	cyan	blue	violet
frequency / 10^{14}Hz	4.3	5.0		6.0	6.7		7.5
wavelength / 10^{-6}m	0.7	0.6		0.5	0.45		0.4

White light can be dispersed, using a triangular **prism**, into the constituent colours.

For more help see Success Guide Physics F p. 46–7 / H p. 42–3

The visible spectrum

- When white light passes into the glass prism, blue light (short wavelength) undergoes a greater change of speed than red light (long wavelength).
- The change in direction depends on the change of speed.

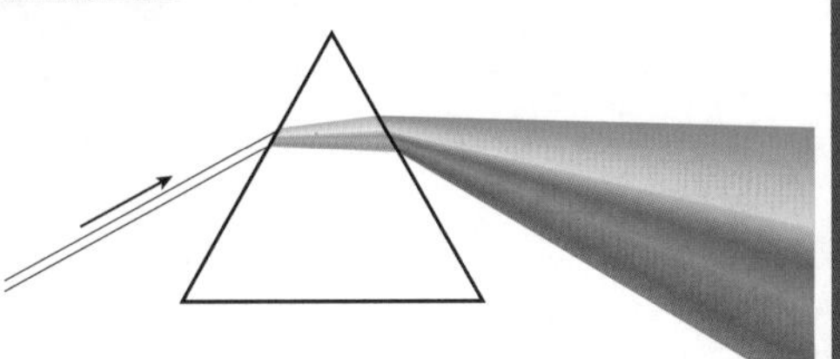

Remember
The order of colours in the spectrum can be remembered using the name ROY G BIV: red, orange, yellow, green, blue, indigo and violet.

Reflection

The diagram shows what happens when light (or another form of electromagnetic radiation) is reflected at a plane surface.

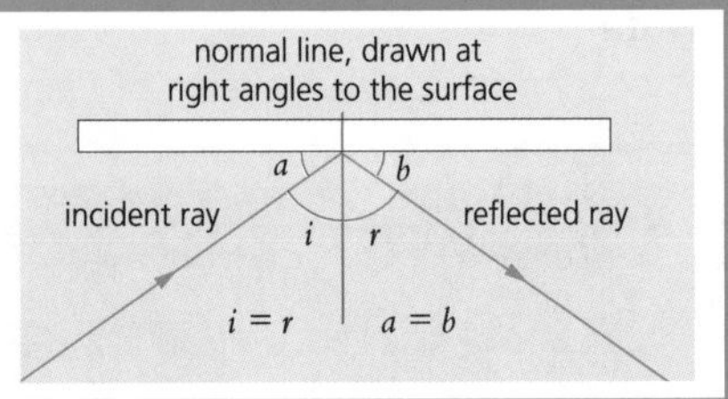

Refraction

The diagram shows how light (or other electromagnetic radiation) passes through a glass block.

TIP!
The ray leaving the glass block is parallel to the ray entering the glass block.

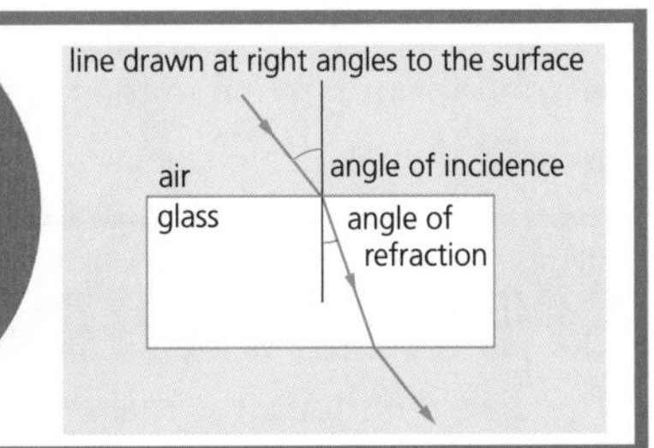

Diffraction

Diffraction is the spreading out of waves when they pass the edge of an obstacle or pass through a gap.

The maximum spreading occurs when the gap is the same size as the wavelength.

Remember
Other types of wave, including sound waves and water waves, undergo diffraction.

For more help see Success Guide Physics F p. 44–5/H p. 38–41

Photosynthesis

Only plants are able to make their own food.
They make this by the process of **photosynthesis**.

The process can be summarised by the equation

$$6CO_2 + 6H_2O \rightarrow C_6H_{12}O_6 + 6O_2$$

carbon dioxide + water → glucose + oxygen.

Photosynthesis takes place in the green leaves of the plant. **Chloroplasts** contain a **pigment** called **chlorophyll** (a **catalyst**). It is an **endothermic** reaction and requires energy from sunlight or a lamp to take place.

Note
Photosynthesis is the reverse of respiration.

What happens to the glucose produced?

The glucose produced in photosynthesis is:

- stored as **insoluble starch** (a carbohydrate polymer)
- used for respiration in the plant
- used to make chlorophyll.

More about the leaf

The diagram shows the main parts of a leaf. Chloroplasts are found near the upper surface of the leaf where they are able to absorb energy from sunlight.

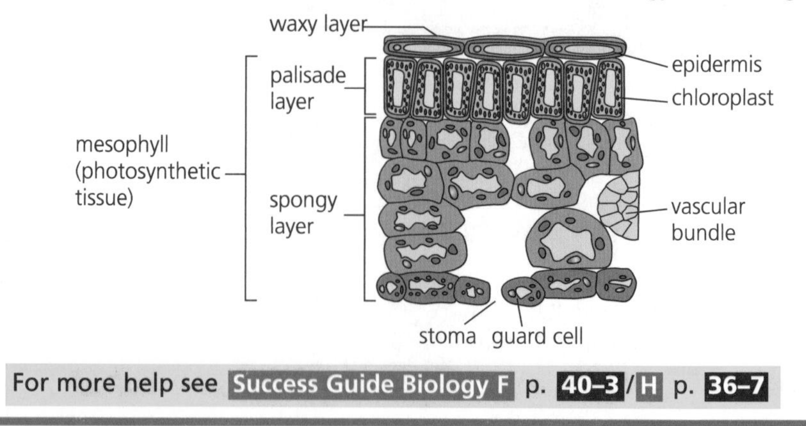

For more help see Success Guide Biology F p. 40–3/H p. 36–7

Photosynthesis with different coloured lights

White light is made up from lights of different wavelengths (colours) The graph shows the relative rates of photosynthesis in different coloured lights.

Because plants are green they reflect much of the green light.

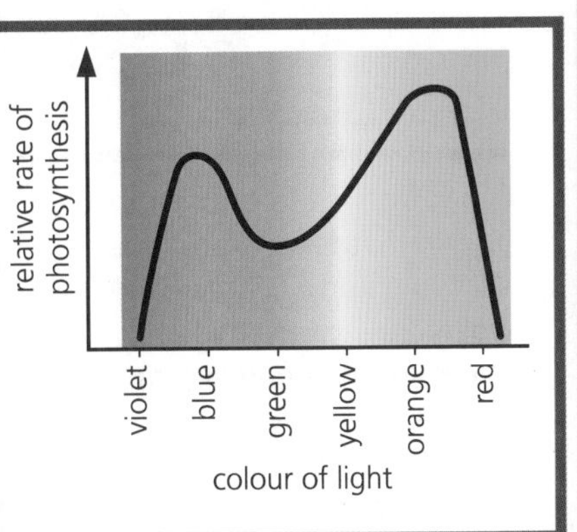

Investigating the rate of photosynthesis

Experiments investigating the rate of photosynthesis of a green plant such an elodea (pondweed) are often followed by tracking the amount of oxygen liberated at regular intervals. This can be done is different ways.

Counting the number of bubbles is unreliable as bubbles are all of different sizes.

Various factors can be studied:

- distance of the lamp from the plant (Remember that light intensity is proportional to the reciprocal of distance squared.)
- levels of carbon dioxide
- temperature.

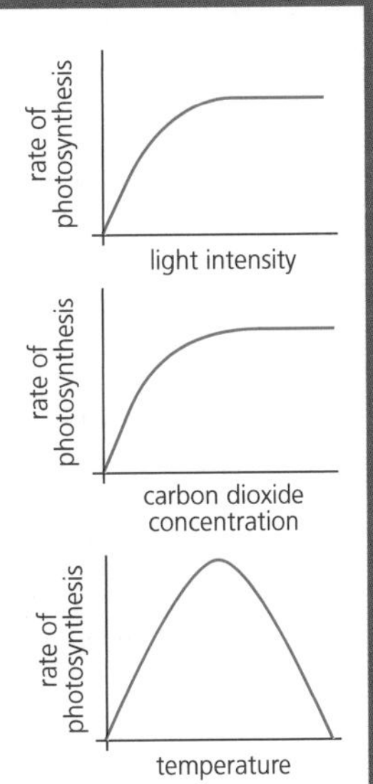

The graphs show how the rate of photosynthesis changes with light intensity, carbon dioxide concentration and temperature.

The rate of photosynthesis increases with increasing light intensity, carbon dioxide concentration and temperature. This continues until a **limiting factor** operates.

In the first graph, the rate is limited when there is a **shortage of carbon dioxide**. Increasing the light intensity then has no effect on the rate of reaction.

In the second graph, the rate is limited by the **light intensity**. Increasing the carbon dioxide ultimately has no effect on the rate.

In graph three, the temperature has a different effect. At low temperatures the rate of photosynthesis is very slow. As temperature rises the rate increases until a certain maximum is reached. At higher temperatures the reaction rate also and then stops.

This curve is typical of **enzyme** processes (see pages 32–3).

Respiration

Respiration is the **oxidation of glucose** that takes place in the cells in the body. Glucose and oxygen are converted into **carbon dioxide** and **water** with the release of a large amount of **energy**.

TIPS!

Students often confuse respiration and breathing. Breathing is a mechanical process that lets air in and out of the lungs. Respiration is the energy releasing process that takes place in the cells.

There are two types of respiration – **aerobic respiration** and **anaerobic respiration**.

Aerobic respiration

Aerobic respiration requires a large amount of oxygen and produces energy more efficiently.

The equation for the reaction is:

$$C_6H_{12}O_6 + 6O_2 \rightarrow 6CO_2 + 6H_2O + \text{energy}$$

glucose oxygen carbon dioxide water

This reaction is **exothermic** (see page 31).
The diagram summarises aerobic respiration.

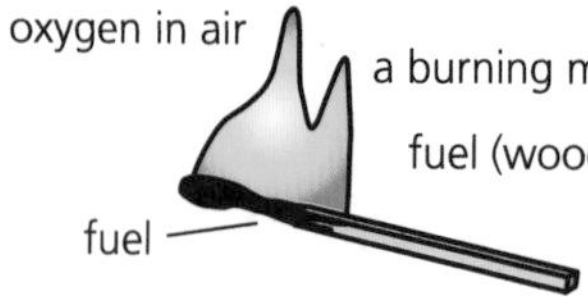

oxygen in air
a burning match
fuel (wood) + oxygen → carbon dioxide + water + energy
fuel

oxygen
carbon dioxide
glucose + oxygen → carbon dioxide + water + energy
water
glucose

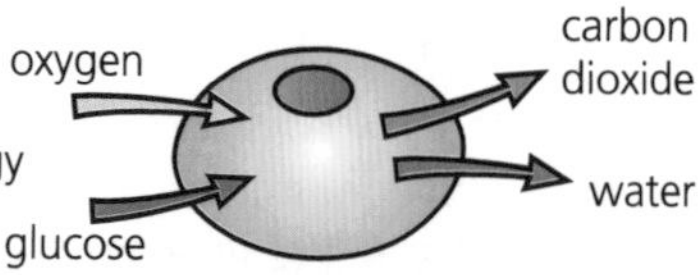

The **waste products**, carbon dioxide and water, are lost by the lungs, the skin or the kidneys.

Energy produced during aerobic respiration has many uses. These include:

- making muscles work
- absorbing molecules against concentration gradients – see **active transport**
- growth and repair of body cells
- building proteins from amino acids
- maintaining body temperature.

For more help see **Success Guide Biology F** p. **20–1**/**H** p. **18–19**

Anaerobic respiration

Anaerobic respiration produces **much less energy** and takes place when there is a **shortage or lack of oxygen**.

An athlete running a long race often gets **muscle fatigue** and **cramp**. This is a consequence of **anaerobic respiration**. The body cannot take in enough oxygen for aerobic respiration to occur. As a result another reaction occurs.

$$C_6H_{12}O_6 \rightarrow 2C_3H_6O_3 + \text{energy}$$

glucose → lactic acid + energy (small amount)

The lactic acid concentration builds up in the muscles and blood. This causes muscle fatigue and cramp.

At the end of the race the **oxygen debt** has to be replaced. The breathing rate and pulse rate remain high, to take in as much oxygen as possible, as the athlete's system continues to oxidise and remove the lactic acid.

Anaerobic respiration also takes place when glucose ferments in the presence of yeast to produce ethanol (alcohol) and carbon dioxide.

Again it is a partial oxidation of glucose and it takes place in the absence of oxygen.

$$C_6H_{12}O_6 \rightarrow 2C_2H_5OH + CO_2 + \text{energy}$$

glucose → lactic acid + energy (small amount)

Fermentation also takes place during bread making when yeast acts on sugar to produce bubbles of carbon dioxide that are trapped inside the dough.

For more help see Success Guide Biology F p. 20–1/H p. 18–19

Energy transfer

There are different ways of transferring energy. These include **conduction**, **convection**, **radiation** and **evaporation**.

Conduction

Metals consist of **tightly packed ions** and free-moving or **delocalised electrons**.

If the end of a piece of metal is heated, the ions **vibrate** faster and faster. The ions cannot move but they can pass on these vibrations to neighbouring ions. Energy can also be passed through the structure by free-moving electrons.

Insulators

Substances other than metals do not conduct heat energy and are called **insulators**.

Materials that do not transmit energy are used to insulate a house. The diagram shows where energy is lost from a typical house.

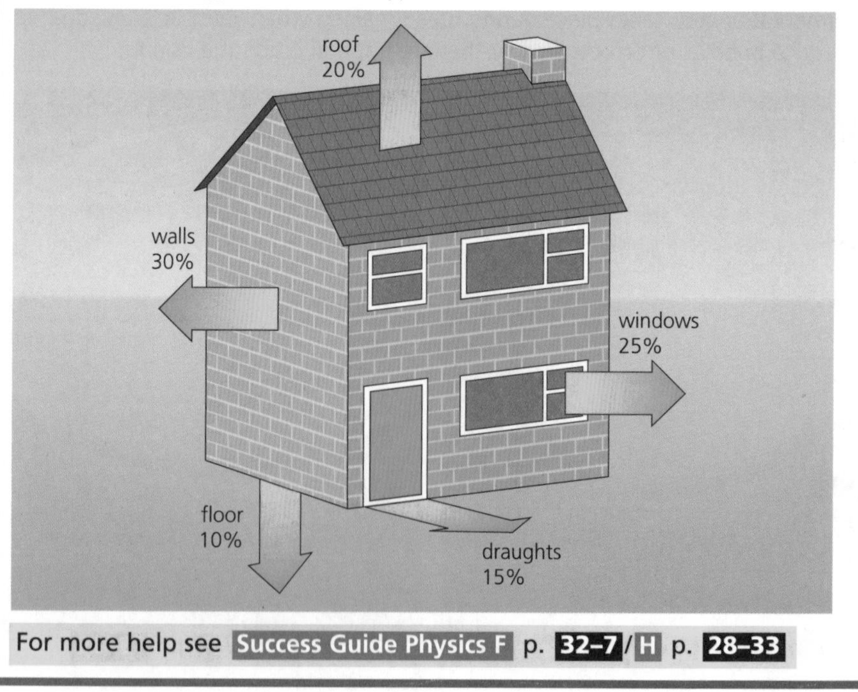

For more help see Success Guide Physics F p. 32–7/H p. 28–33

Convection

The diagram shows how **convection currents** can heat a room. Convection currents are circular movements of a fluid (liquid or gas).

TIP!

It is not enough simply to write that hot air rises. You must explain it in terms of differences in density.

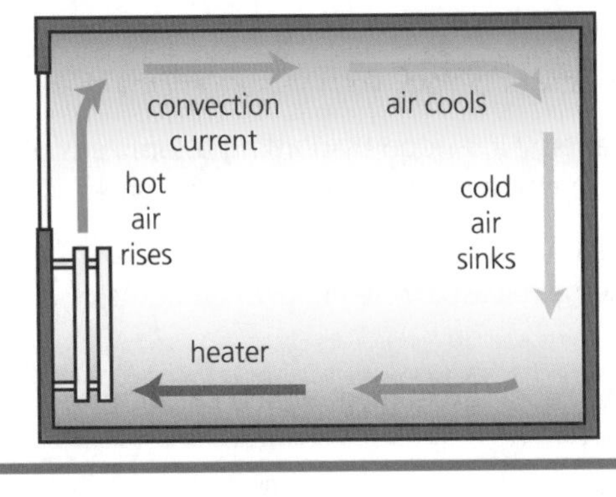

Hot air rises because it is **less dense** than cold air. Cold air falls because it is **more dense** than hot air.

Radiation

Energy transfer by **radiation** is most efficient because there are **no particles** needed. It is a **direct transfer**. The energy from the Sun is radiated to reach the Earth through a **vacuum**.

When electromagnetic radiation strikes an object, that energy can be **absorbed** or **reflected**.

Objects that are **dark and have rough surfaces** absorb most of the energy. Objects that are **light-coloured and shiny** reflect most of the energy.

Objects that are good absorbers are also good emitters of energy. Under the same conditions, a dark, dull object will cool faster than a light-coloured, shiny one.

Evaporation

Evaporation is the process where a liquid turns into a vapour. It can take place at all temperatures.

Evaporation is accompanied by **cooling**.

In a refrigerator, the liquid is made to evaporate, taking energy from the inside of the refrigerator. At the back of the refrigerator there is a **condenser** where the **vapour turns back to a liquid**. This feels hot because energy is released when condensing occurs.

For more help see **Success Guide Physics F** p. **32–7**/**H** p. **28–33**

Measuring heat changes

In many experiments in physics, chemistry or biology, you measure energy changes.

In biology you might measure the energy in a food by burning a sample of the food and measuring the change in temperature of a fixed mass of water.

In chemistry you might measure the energy released when known amounts of acid and alkali are mixed or the energy released when known amounts of a fuel are burned.

These experiments use **calorimetry**, i.e. an experiment with a **calorimeter**.

Burning alcohols

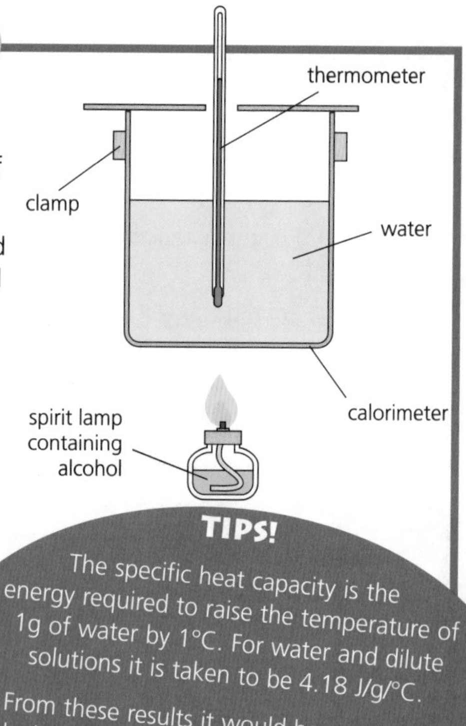

Alcohols burn in air to release carbon dioxide, water and energy. An alcohol in a spirit lamp is used to heat 100g of water in a calorimeter.

The mass of alcohol used can be found by weighing the spirit lamp before and after the experiment.

If the temperature rises from 20°C to 35°C, the temperature rise is 15°C.

The energy released can be calculated using the equation:

energy released
= mass of water × specific heat capacity × temperature rise
= $ms\theta$
= $100 \times 4.18 \times 15$
= 6270 J or 6.27 kJ

A stirrer might help to ensure that the temperature is the same throughout the solution.

TIPS!

The specific heat capacity is the energy required to raise the temperature of 1g of water by 1°C. For water and dilute solutions it is taken to be 4.18 J/g/°C.

From these results it would be possible to calculate the energy released when 1 formula mass of the alcohol is burned. This is the **heat of combustion**.

Comparing results with the data book

In an experiment such as this the result obtained is much lower than the value in the data book. This is because much of the energy released is lost to the surroundings and is used to heat the calorimeter and thermometer.

Improvements could include:

- using draught shields
- keeping the temperature rise as low as possible
- using apparatus that reduces energy losses to the surroundings (see diagram).

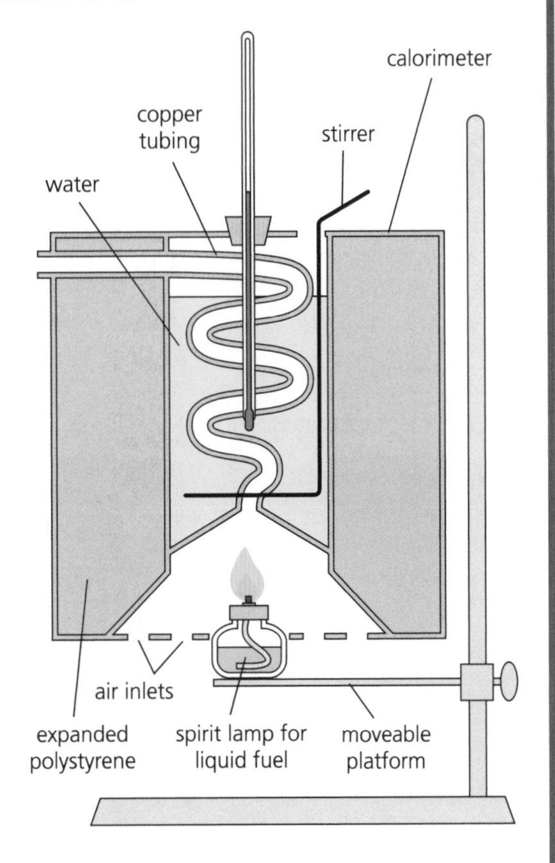

Exothermic reactions

Combustion reactions, like respiration, release energy and are called an **exothermic reaction**.

The energy level diagram summarises the changes during an exothermic reaction.

reactants

↓ **energy released**

products

For more help see Success Guide Chemistry F p. 78/H p. 89

Mitosis and meiosis

There are two methods of cell division – **mitosis** and **meiosis**.

Mitosis

The process of mitosis occurs when a cell divides to produce **two daughter cells** each **identical** to the original parent cell. The process is summarised in the diagram. There are 46 chromosomes in each body cell – in 23 pairs. In this diagram only four (two pairs) are shown for simplicity.

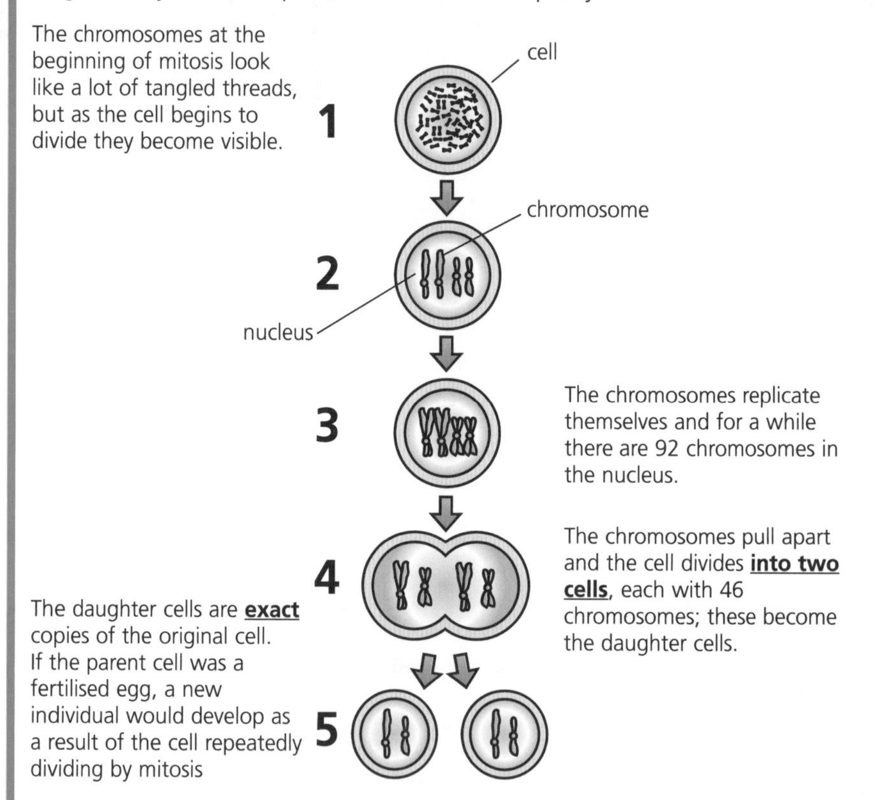

Asexual reproduction uses only one parent. For example, a new plant can be grown from a cutting of an existing plant or an organism can be **cloned** or copied using cells from one organism. The offspring are identical to the parent because the genes are identical.

For more help see **Success Guide Biology H** p. **64–5**

Meiosis

In meiosis, **gametes** (or **sex cells**) are produced in the ovary of the female and the testes of the male.

These gametes contain only half the number of chromosomes compared with body cells. Human body cells contain 46 chromosomes but gametes contain only 23. The diagram summarises the process of meiosis. Again, only four chromosomes (two pairs) are shown.

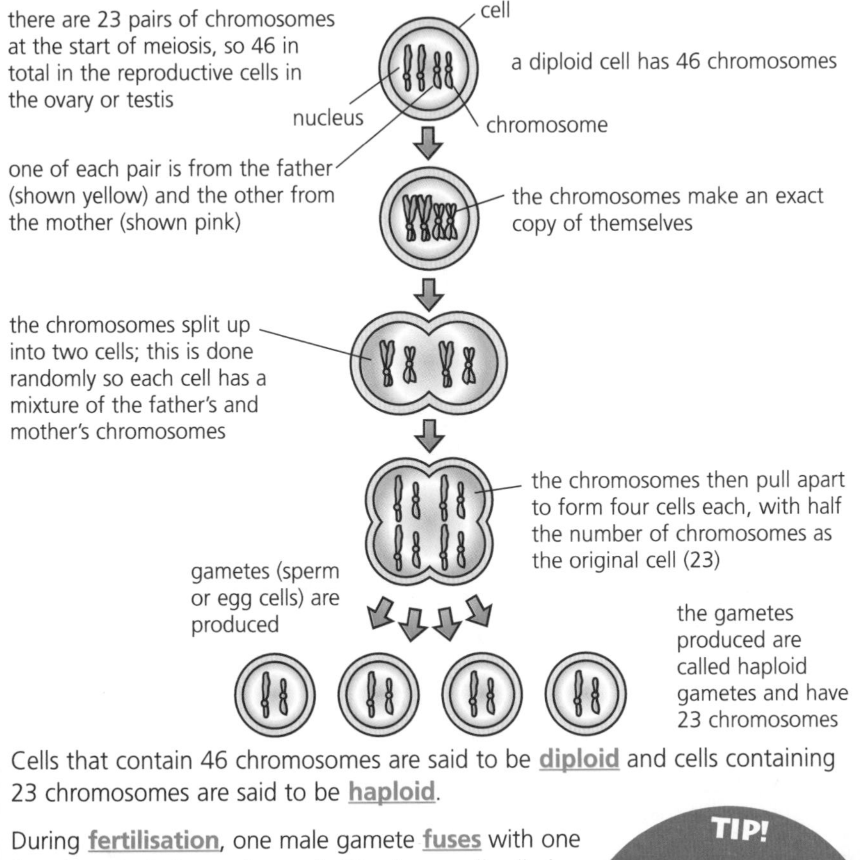

Cells that contain 46 chromosomes are said to be **diploid** and cells containing 23 chromosomes are said to be **haploid**.

During **fertilisation**, one male gamete **fuses** with one female gamete to produce a fertilised egg cell called a **zygote**. The zygote again contains 46 chromosomes or 23 pairs. Half of these chromosomes have come from the male and half from the female.

Meiosis and fertilisation lead to **variation** because the offspring inherit a combination of genes from each parent.

TIP!

Make sure you spell these technical terms correctly. A word written as 'meitosis' has to be marked incorrect because it is not clear whether meiosis or mitosis is meant.

For more help see **Success Guide Biology H** p. **66–7**

Selective breeding and genetic engineering

Selective breeding

A farmer wants his sheep to have fleeces with thicker wool. The diagram summarises how he can do this using selective breeding.

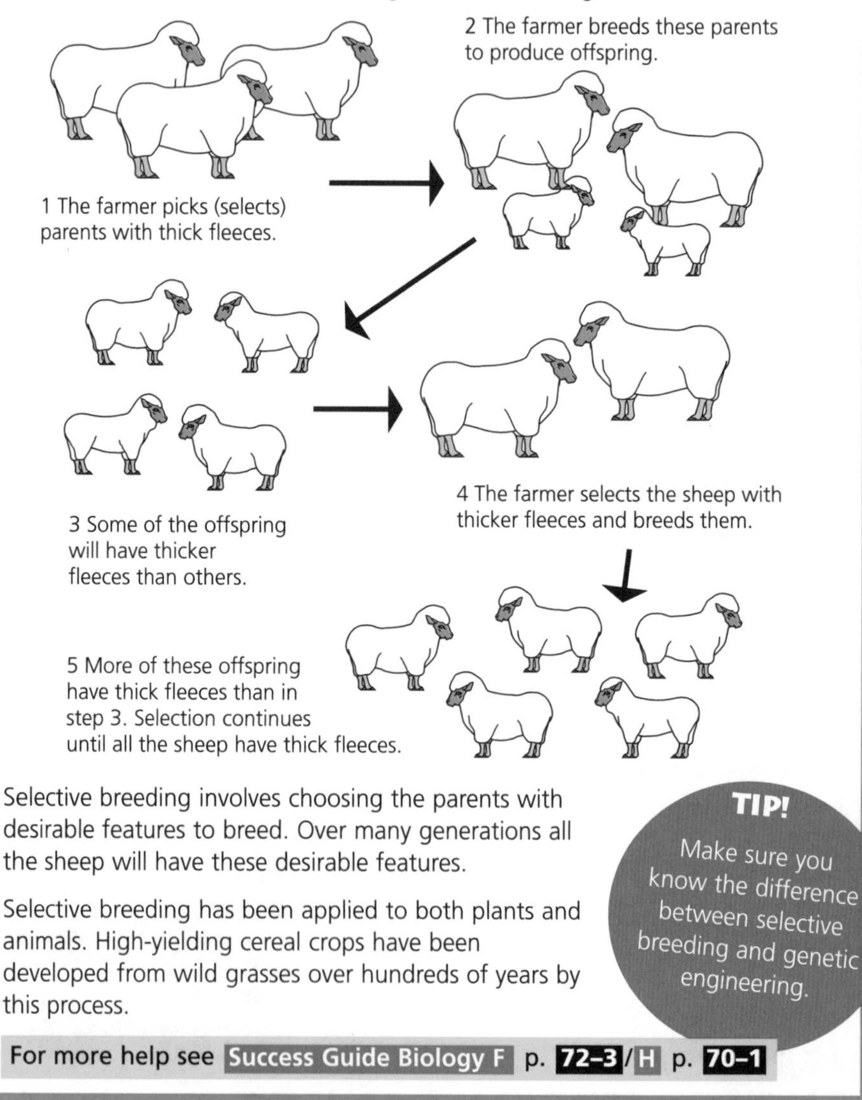

Selective breeding involves choosing the parents with desirable features to breed. Over many generations all the sheep will have these desirable features.

Selective breeding has been applied to both plants and animals. High-yielding cereal crops have been developed from wild grasses over hundreds of years by this process.

TIP! Make sure you know the difference between selective breeding and genetic engineering.

For more help see Success Guide Biology F p. 72–3/H p. 70–1

Genetic engineering

Genetic engineering is the process by which genes from one organism are removed and transferred into the cells of another organism.

The diagram shows how a gene from plant A can be inserted into plant B.

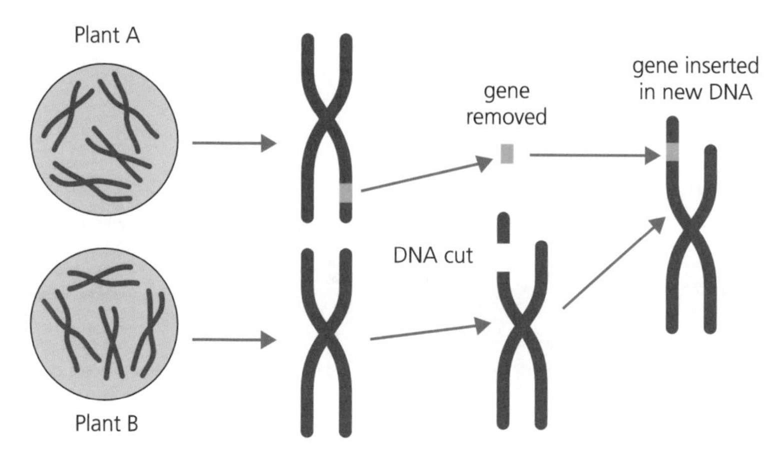

There are some potential benefits of genetic engineering.

- People with inherited diseases like cystic fibrosis may be able to be treated in the future by replacing the gene that causes the disease in their cells.
- Genetic engineering already produces insulin, for diabetic patients.
- Weed killer-resistant plants such as maize or wheat could be grown. Farmers could spray to kill the weeds without killing the crop.

However, many people have doubts about genetic engineering.

- Some religious beliefs rule out this sort of work.
- Some people are doubtful about the potential side effects of the process.
- Others worry about how the technique might be used by commercial companies. In many parts of the world farmers collect seed from one year and save it for the next. If non-germinating seeds were developed by genetic engineering, those farmers would have to buy new seed each year.

For more help see **Success Guide Biology F** p. **66–7**/**H** p. **60–1**

Homeostasis

Homeostasis is the mechanism used by the body to maintain a **constant internal environment**.

Examples of homeostasis are:

- keeping the body temperature constant
- controlling levels of water in the body
- controlling blood sugar levels.

Keeping the body temperature constant

The **skin** is the most important organ for maintaining constant body temperature. The diagram shows the structure of the skin.

outer layer
made of dead flat cells that flake off. They are replaced by cells from below and provide a barrier to the outside environment

hair root
the living part of the hair contained in the hair follicle

hair

sweat pore

sweat glands
produce liquid that is secreted at the surface to cool the body down

sebaceous glands
produce sebum, an oily substance that makes the skin waterproof and supple

fat layer
for insulation

blood vessel

blood capilliary

Body temperature is monitored by particular cells in the **brain** (in the **hypothalamus**). These are sensitive to small temperature changes. If activated by changes in temperature they set up mechanisms to restore conditions.

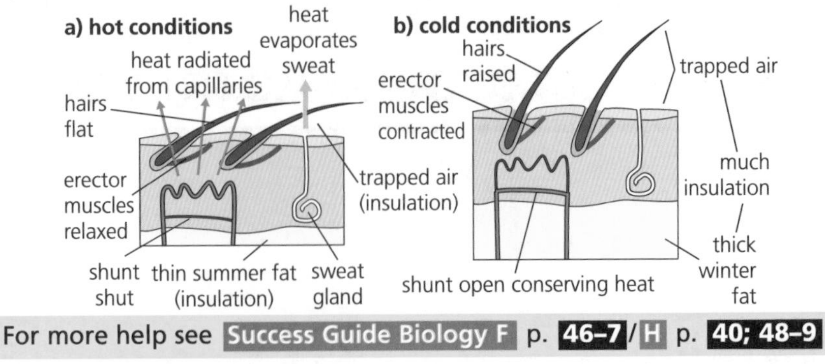

For more help see Success Guide Biology F p. 46–7 / H p. 40; 48–9

Controlling water levels

The control of water in the body is called **osmoregulation**. It is monitored by the **pituitary gland** in the **brain**. The brain releases an **anti-diuretic hormone** (ADH) depending upon the concentration of water in the blood.

If the concentration of water is low	If the concentration of water is high
because you have drunk little water	because you have drunk a lot of water
no ADH is produced and kidneys do not reabsorb water.	ADH is produced and water is reabsorbed by the kidneys.
A lot of dilute urine is produced.	Little concentrated urine is produced.
Blood returns to normal.	

Controlling blood sugar levels

Hormones are chemicals that travel around the body in the blood and target the body's organs. The hormone **insulin** is produced by the **pancreas**. Its job is to keep the levels of glucose in the blood constant. It causes the **liver** to store glucose (**glycogen**) until it is needed.

A shortage of insulin causes blood glucose levels to be too high. This condition is called **diabetes** and, if left untreated, diabetes can lead to weight loss and even death.

The diagram shows how the pancreas controls the blood sugar level by a process called **negative feedback**.

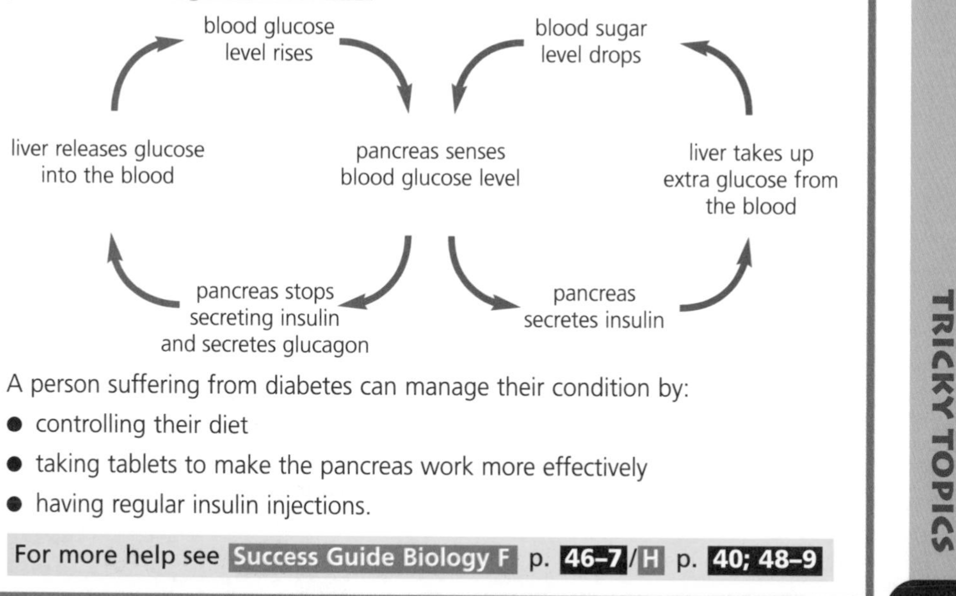

A person suffering from diabetes can manage their condition by:

- controlling their diet
- taking tablets to make the pancreas work more effectively
- having regular insulin injections.

For more help see Success Guide Biology F p. 46–7/H p. 40; 48–9

Acceleration

Acceleration is the change of velocity per unit time.

$$\text{acceleration} = \frac{\text{increase in velocity}}{\text{time}}$$

Example

An aircraft speeds up from 10 m/s to 60 m/s in 10 seconds.

The acceleration = $\frac{(60-10)}{10}$ = 5 m/s^2.

TIPS!

A frequent mistake is to give the units of acceleration as m/s/s. It should be m/s^2

Getting acceleration from a graph

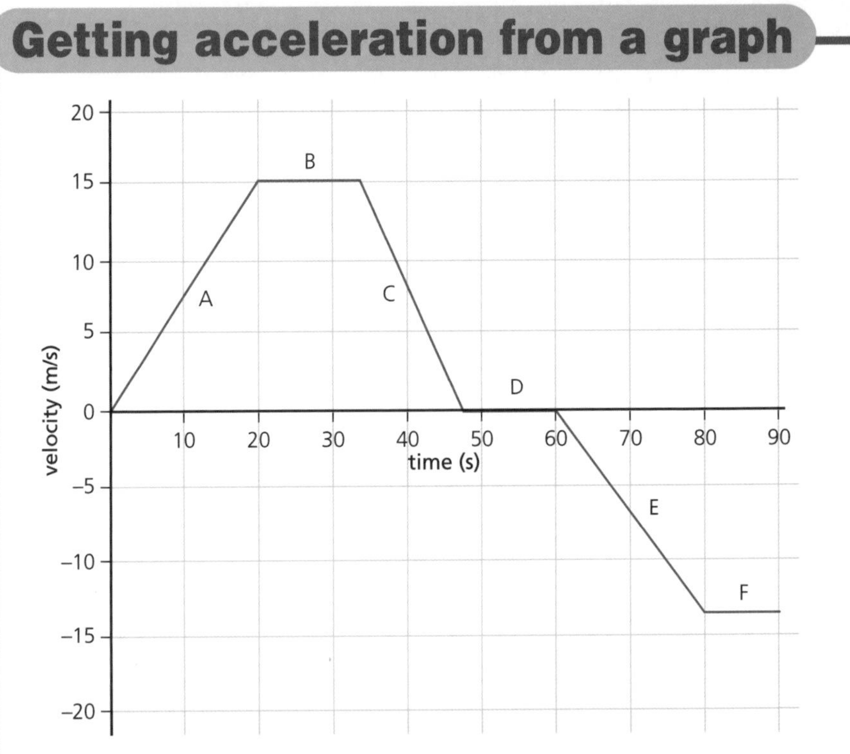

The graph shows velocity against time for a car on a journey. The acceleration at any time can be found by calculating the gradient of the graph: at A it is 0.75 m/s^s and at E it is –0.65 m/s^2. The positive sign shows the car is speeding up and the negative sign slowing down (**deceleration**).

For more help see Success Guide Physics F p. 8–11/H p. 4–9

Measuring acceleration

The diagram shows a toy car being pulled down a ramp.

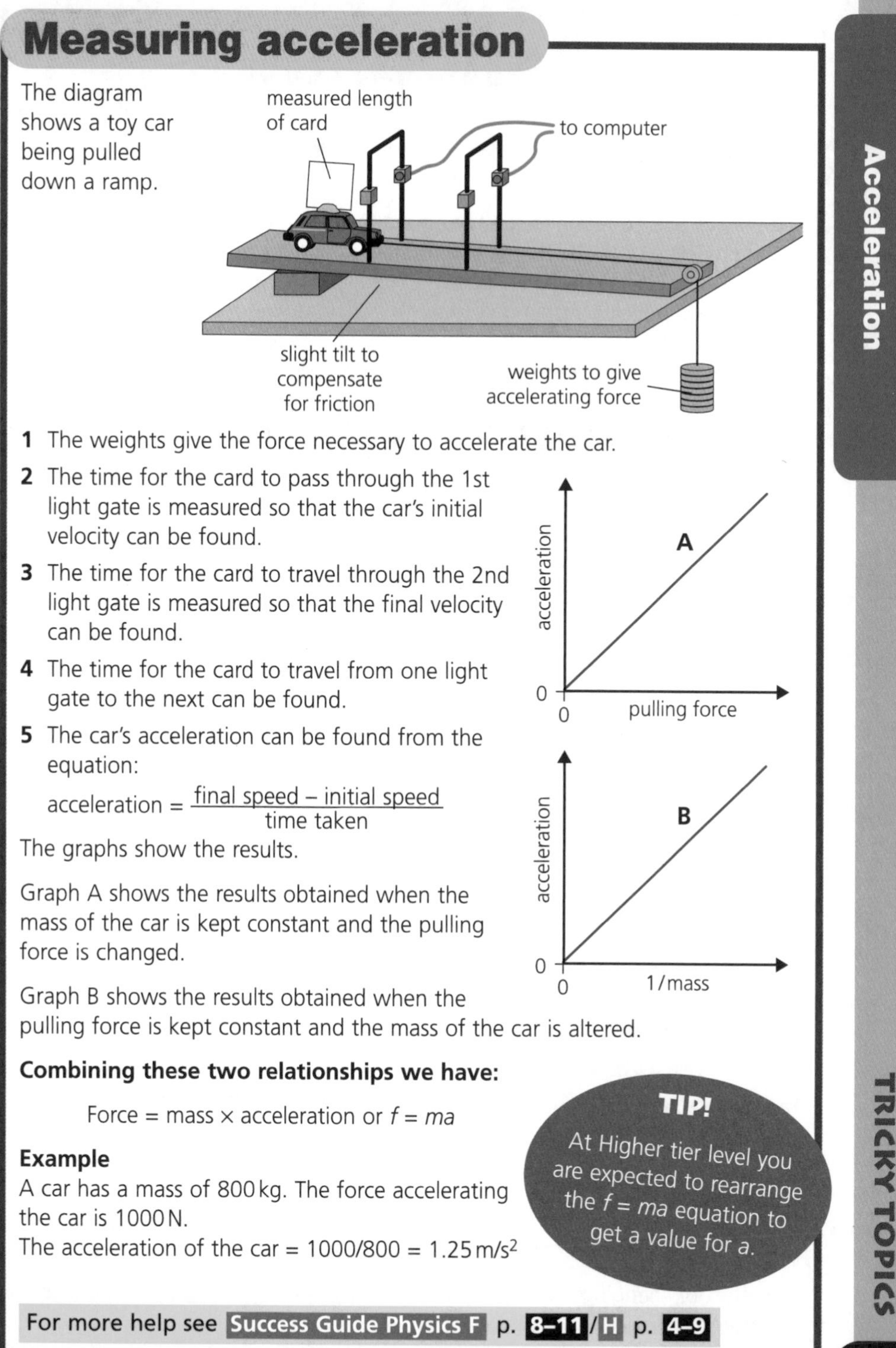

1 The weights give the force necessary to accelerate the car.

2 The time for the card to pass through the 1st light gate is measured so that the car's initial velocity can be found.

3 The time for the card to travel through the 2nd light gate is measured so that the final velocity can be found.

4 The time for the card to travel from one light gate to the next can be found.

5 The car's acceleration can be found from the equation:

$$\text{acceleration} = \frac{\text{final speed} - \text{initial speed}}{\text{time taken}}$$

The graphs show the results.

Graph A shows the results obtained when the mass of the car is kept constant and the pulling force is changed.

Graph B shows the results obtained when the pulling force is kept constant and the mass of the car is altered.

Combining these two relationships we have:

Force = mass × acceleration or $f = ma$

Example
A car has a mass of 800 kg. The force accelerating the car is 1000 N.
The acceleration of the car = 1000/800 = 1.25 m/s^2

TIP!
At Higher tier level you are expected to rearrange the $f = ma$ equation to get a value for a.

For more help see Success Guide Physics F p. 8–11/H p. 4–9

Pendulum

A pendulum consists of a mass attached to a string or light rod.

The mass is able to rotate in the Earth's gravitational field round a pivot that is not at its centre of mass. The pendulum was studied by the Italian scientist Galileo Galilei (1564–1642). He became interested in the pendulum when watching a chandelier suspended from the ceiling during services in the cathedral at Pisa.

The movement of a pendulum

A simple pendulum is shown in the diagram.

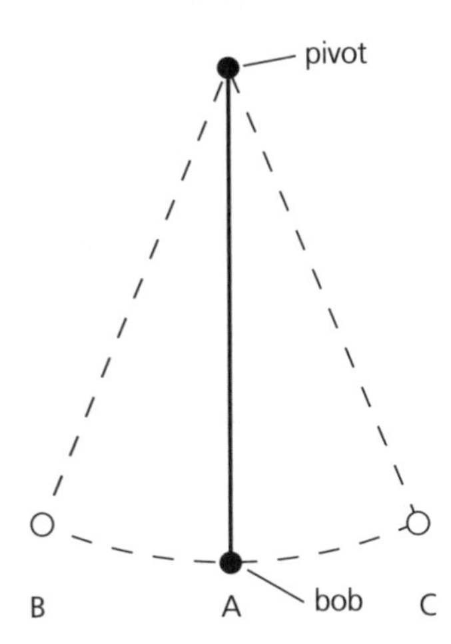

At point A where the mass, called the **bob**, is vertically below the pivot, the bob is moving fastest. As it moves between A and B the bob slows down, until at B the bob is stationary. Here it changes direction, speeding up as it returns from B to A.

The time for one swing (sometimes called the period) is the time it takes for the pendulum to move from B to C and back to B again.

In simple experiments students might study:

The effect of changing the distance between the bob and the pivot. **Increasing the distance between the bob and the pivot increases the period**.

The effect of changing the mass of the bob. **Changing the mass of the bob has no effect on the period**.

Energy changes during the swing of a pendulum

At point A the bob is moving fastest and so possesses **maximum kinetic energy** but **zero potential energy**. As it moves from A to B, it loses kinetic energy (or slows down) and gains potential energy. At B it has **zero kinetic energy** but **maximum potential energy**.

Looking at a pendulum in more detail

Students studying a pendulum at Higher tier level might use the equation:

$$T = 2\pi\sqrt{\frac{l}{g}}$$

where T is the time for one swing (period)
l is the length (i.e. distance from pivot to bob)
g is gravitational field strength 10N/kg

After a series of experiments, a graph of T^2 against l should be a straight line. A value for g can be worked out from the gradient of the graph.

This equation is only true for small amplitudes, i.e. when the distance A to B is small. Students could study the effect of changing the amplitude.

What causes the pendulum to slow down?

A pendulum will gradually slow down during an experiment. This can be caused by:

1 air resistance as the pendulum moves through the air

2 friction in the pivot.

Measuring the period

Measuring the time for one complete swing (or period) is very difficult as the time will be short. It is better to time 10 swings and then work out the time for one swing by dividing by 10.

Using a light gate at A is probably a better solution but remember that the bob will pass through the light gate twice in one period.

Electromagnetic induction

When the **magnetic field** through a coil **changes**, it causes a **voltage** across the terminals of the coil.

This is called magnetic induction.

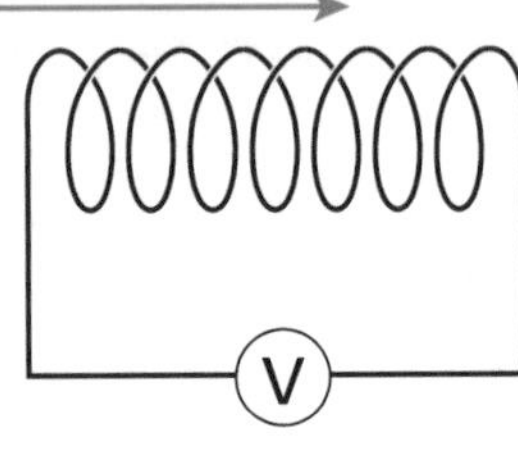

The size of the induced voltage is increased by:

- increasing the number of turns on the coil
- moving the magnet faster
- increasing the strength of the magnetic field.

The direction of the induced voltage can be reversed by:

- reversing the direction of the movement
- reversing the poles of the magnet.

TIPS!

The most common mistake is to forget that a voltage is only induced if movement is involved. No induced voltage is shown wher magnet is stationary within th coil. A similar effect occurs if a coil is moved round a stationary magnet.

Using electromagnetic induction

Electromagnetic induction is used:

1 in an **a.c. generator**. A **coil** of wire is **rotated** within a **magnetic field**. This causes a voltage across the terminals. An a.c. generator is used to provide electricity in a car.

2 in a bicycle **dynamo**. This uses a **rotating magnet**.

The faster the bicycle goes the greater the induced voltage and the brighter the bike's lamps.

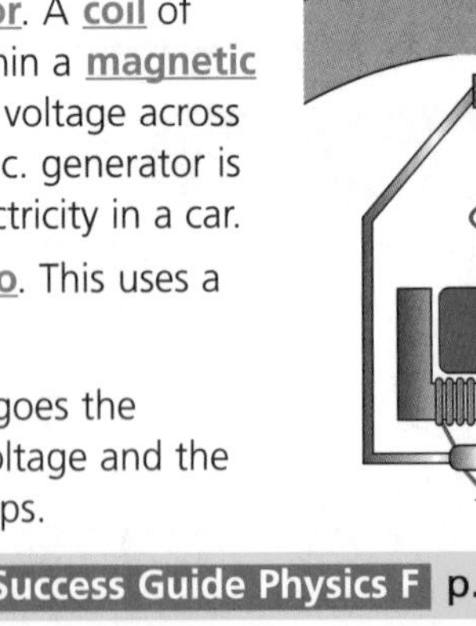

For more help see Success Guide Physics F p. 72–3/H p. 70–1

The transformer

A transformer is used to **change a.c. voltages**. The voltage may be:

increased by a **step-up transformer**, for example, before electricity is transmitted around the country in the National Grid

decreased by a **step-down transformer**, for example, to reduce the voltage of mains electricity for a train set.

A transformer consists of **two coils of wire** wrapped around a **soft iron core**.

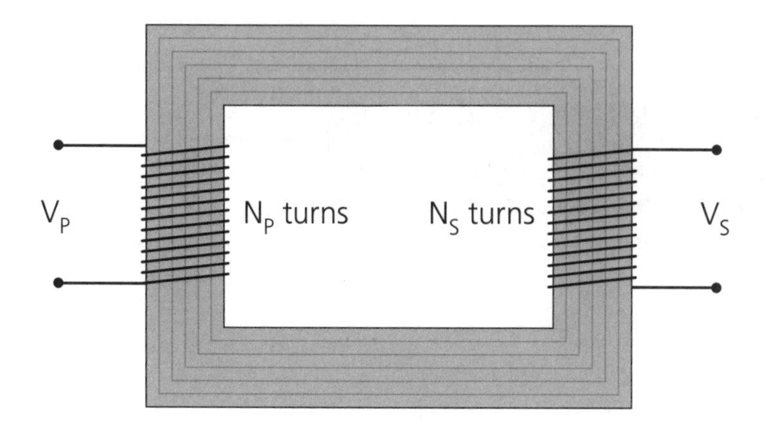

The relationship between the input (primary) and the output (secondary) voltages of a transformer is:

$$\frac{\text{primary voltage}}{\text{secondary voltage}} = \frac{\text{number of primary turns}}{\text{number of secondary turns}}$$

$$\frac{V_P}{V_s} = \frac{n_P}{n_s}$$

Example

A transformer has an input voltage of 240V and 1000 turns on the primary coil. How many turns are needed on the secondary coil to make the output voltage 12V?

$$n_s = \frac{n_P V_S}{V_P}$$

$$= \frac{1000 \times 12}{240}$$

$$= 50 \text{ turns}$$

TIP!

This is a step-down transformer and there are always more turns on the primary coil than on the secondary.

For more help see Success Guide Physics F p. 74–5/H p. 72–3

Ideas and evidence

All GCSE science papers have 'ideas and evidence' questions. They account for 5% of the specification, or 6% of the written papers. They do not appear in coursework.

These questions are not usually identified as 'ideas and evidence' on the paper but they will give you additional opportunities to interpret data and to propose a scientific argument.

Science is not just about learning facts

In 'ideas and evidence' you might be expected to deal with historical questions, such as the development of ideas about:

- the circulation of the blood round the body
- the Periodic Table (see page 8)
- plate tectonics and the structure of the Earth.

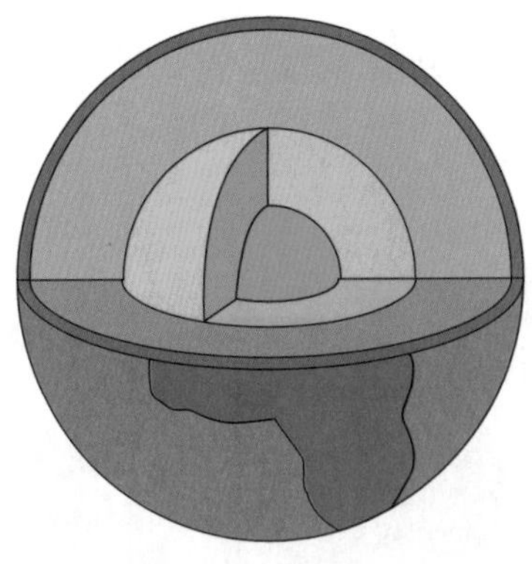

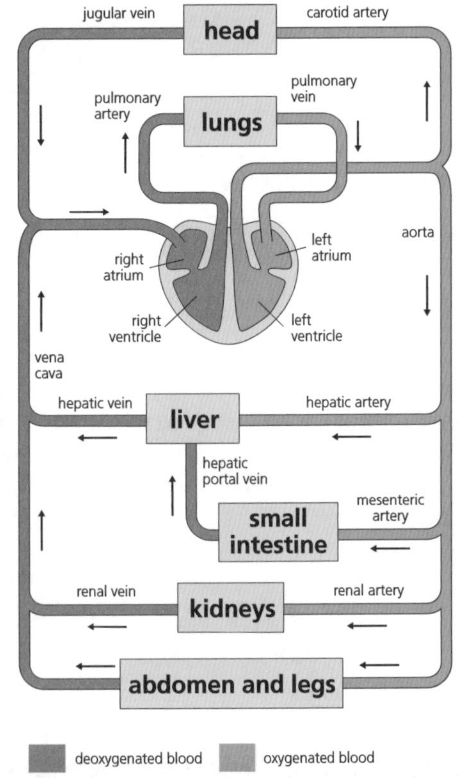

The questions do not usually require you to recall a lot of information but to interpret and use the information given in the question.

There are four strands to 'ideas and evidence'.

1 How scientific ideas are presented, evaluated and disseminated.

2 How scientific controversies can arise from different ways of interpreting empirical evidence.

3 Ways in which scientific work may be affected by the context in which it takes place and how different contexts may affect whether or not ideas are accepted.

4 The power and limitations of science in addressing industrial, social and environmental questions, including the kind of questions science can and cannot answer, uncertainties in scientific knowledge and ethical issues.

You cannot expect questions from all categories to appear on any one examination paper but over a few years all will be covered several times.

Good general knowledge, improved by reading articles in newspapers, magazines or the Internet, will help you answer these questions.

When you are putting forward an argument, make it clear what your point of view is and then use your scientific knowledge to support your argument. Make sure the examiner reading your answer is in no doubt whether your opinion is for or against.

Try to include as much science as possible in these answers. Many candidates write a great deal but without including sufficient science and frequently repeating the same points.

A close look at Sc1.2

Sc1.2 is the practical coursework you have to do in school or college as part of your GCSE.

The table gives the percentage of marks allocated to each attainment target for Double Award science, Single Award science and single sciences.

	Sc1.1 Ideas and evidence	Sc1.2 Practical assessment	Sc2 Biology	Sc3 Chemistry	Sc4 Physics
Science DA	5	20	25	25	25
Science SA	5	20	25	25	25
Biology	5	20	75		
Chemistry	5	20		75	
Physics	5	20			75

During years 10 and 11 there will be opportunities for your teacher to assess your performance in Sc1.2 in four **Skill Areas**.

- Skill Area P – Planning experiments
- Skill Area O – Obtaining evidence
- Skill Area A – Analysing evidence and drawing conclusions
- Skill Area E – Evaluating the evidence.

There is no limit to the number of times you can be assessed. Your two best marks in each Skill Area for will count for Double Award, within certain restrictions.

For Single Award and separate biology, chemistry and physics only one mark from each Skill Area counts.

Double Award

You must have work from at least **two** of biology, chemistry and physics and at least one mark must come from a whole investigation. You must not have more than four pieces of work in total.

For Single Award and single sciences

You only need work from **one** of biology, chemistry and physics and one mark must come from a whole investigation. You must not have more than two pieces of work.

Using ICT

You can use ICT to write your reports or draw graphs. You can also use other aspects of ICT. However, you must always do some lab-based practical work.

SIMULATIONS

Instead of doing practical work it is possible to buy simulations of experiments and use them to plan and obtain data. This is quite acceptable but the software sometimes has certain restrictions that may limit the highest mark you can achieve.

SPREADSHEETS

It is possible to set up a spreadsheet to process data: for example, to do routine calculations of resistance.

INTERNET SEARCHES

You can carry out a search of the Internet to find data to analyse and evaluate. It is important that you plan carefully what you are going to do and record the addresses of any websites you use.

MODELLING

You can use your computer to model theoretical situations that you cannot create easily in the laboratory.

Using ICT is not compulsory but it can have some advantages for Sc1.2.

Working through Sc1 criteria

Marks	P: Planning experiments Planning my work	O: Obtaining evidence Collecting my data
2	I can plan something to investigate.	I can work safely to collect my evidence.
4	+ I know that my plan will allow me to collect evidence I can use. I know what equipment I need to collect my evidence.	+ I can collect enough evidence to give me an answer. I can record my evidence carefully.
6	+ Using scientific ideas I can decide what are the most important factors. I can say clearly what readings I will need to take and how many of them I will need.	+ I can collect evidence carefully and accurately and record it clearly, showing the correct units. I can check the quality of my evidence by taking repeats or making extra checks.
8	+ I can use detailed scientific ideas in my plan to help me to get reliable results. I have given details of any preliminary work which has helped me in my final planning and shown how it is used.	+ I can use equipment skilfully to collect sufficient high-quality results. I use sophisticated apparatus in my experiments.

A: Analysing evidence and drawing conclusions Making my conclusions	Marks	E: Evaluating the evidence Evaluating my work
I can show what I have found out.	**2**	I can say something about how well my plan has worked.
+ I can use diagrams, charts or simple graphs to help me show if there is a pattern in what I have found out.	**4**	**+** I can say how accurate my results are and if my experiments can be improved. I can recognise results that do not fit the pattern.
+ I can use my scientific knowledge and the information from my diagrams, charts or graphs to help me make a conclusion.	**6**	**+** I can explain clearly if my evidence is reliable enough to support my conclusion. I can suggest how to get extra useful evidence in this activity.
+ I can use detailed scientific ideas to explain my conclusion. I can say how my results link back to my planning.		

Skill Area P

The first task in a whole investigation is to complete a planning exercise. It is impossible to have a successful investigation unless there is thorough planning incorporating science knowledge and understanding. In fact, it is impossible to score six marks without fully involving your scientific knowledge and understanding and for eight marks this must be of grade A standard.

Before you start an investigation, research the topic thoroughly.

Variables

It is important that you understand the term **variable**.

A variable is a factor that can be measured or controlled.

A variable you can change is called an **independent variable**. A variable over which you have no control is called a **dependent variable**.

Example
If you are carrying out a rate of reaction experiment where you are altering the temperature and measuring the time for the reaction, temperature is the dependent variable and time is the independent variable.

In an investigation, some variables are **key variables** as they are particularly important. You should identify key variables and decide how to control them.

Sometimes, especially in biology, situations can be so complex, with so many variables operating, it is very difficult to alter a single variable.

If you manage to keep all variables fixed and change only one, you can decide what effect that particular variable has. This is called **fair testing**.

TIPS!
Fair testing is not doing something more often or more carefully.

Variables can be classified as **discrete** or **continuous**.

DISCRETE VARIABLE

A discrete variable has only a certain number of possibilities but a continuous variable can have all values.

Example
In a simple experiment to find the number of indigestion tablets required to neutralise a given volume of acid, the answer must be a whole number (1, 2, 3, etc.). There are no intermediate values so this is a **discrete** variable.

Shoe size is another discrete variable.

Variables – continued

CONTINUOUS VARIABLE

A continuous variable can have any value within a range.

Example
The heights of a group of people are continuous and do not fall into distinct categories.

You are only expected to study the effect of one key variable in detail. There is no advantage in trying to study more than one variable.

Making a prediction

When you are planning you should attempt to make a prediction using your knowledge and understanding. A prediction without any science is only a guess.

You are going to collect and process data to check whether the prediction you made is correct.

Choosing apparatus

For P4b you are expected to choose suitable apparatus for the investigation. In some schools and colleges, the equipment needed is sometimes already provided and so choice is not possible. You can still justify the choice of equipment, however.

In most schools and colleges three types of thermometer are available:

- –10°C to 110°C with 1°C divisions
- 0°C to 50°C with 0.1°C divisions
- 0°C to 360°C with 2°C divisions.

In a particular investigation using a thermometer you could explain which of these three thermometers would be the best and why.

Preliminary work

A practical investigation takes a considerable amount of time to do well. It is a good idea to do some preliminary work before you start the investigation.

You cannot score P8b unless you have completed suitable preliminary work, recorded the results and shown how this work informed your plan.

Too often students fail to write up preliminary results in the same way that they write 'ordinary' results.

Preliminary work for an experiment

Example
A rate of reaction experiment with dilute hydrochloric acid and calcium carbonate.

Carry out a preliminary experiment to find what concentration of hydrochloric acid should be used.

Choose a mass of calcium carbonate and a temperature. Then try some different concentrations of hydrochloric acid. You do not want the reaction to be too fast or too slow. If you are collecting the volume of carbon dioxide in a $100\,cm^3$ measuring cylinder, a final volume of $90\,cm^3$ would be ideal.

From the results of your experiments you should be able to plan how to carry out the experiment well.

There are computer-simulation programs that can be used to give you the same information. They are quite acceptable in your preliminary work.

TIP!

If you use a computer in your planning, make sure you clearly reference any source you have used. This enables your teacher or a moderator to check.

Choosing a suitable range of observations and measurements

One important task you have in planning is to decide the number and range of measurements you will make.

There is no correct answer. If taking the measurements is a lengthy and complicated process you may have to limit the number of results you take.

As a general rule:

- try to get as wide a range as possible
- try to take readings at five different points
- repeat the results two or (even better) three times.

Example
You are studying the rate of reaction between sodium thiosulphate and dilute hydrochloric acid, at different temperatures.

It is sensible to start at room temperature and finish at a maximum of about 55°C. From your scientific knowledge you will know that sulphur dioxide is also produced and this becomes a problem above 55°C.

A suitable set of values might be 20°C, 28°C, 35°C, 44°C and 53°C.

Avoid 20°C, 25°C, 30°C, 35°C and 55°C.

Here the range lies more at the lower end.

In an enzyme experiment different temperatures are used. If you choose 20°C, 25°C, 30°C, 75°C and 85°C, your results will be useless. At the temperatures chosen there will be little enzyme activity. The important range through which results will be needed is 30–55°C Your scientific knowledge will help you decide this.

Before finishing your plan, give it a final read through.

- Has it got real scientific information within it, for example particle model, collision theory, lock and key model, etc?
- Is there sufficient detail in your plan for you to be able to carry out the investigation without any help?

Skill Area O

This Skill Area is about collecting and displaying observations and measurements from your investigation.

Remember these points.

1 Record your observations and measurements as soon as you make them. Don't rely on remembering them! A table is the best way to record them.
2 Design your table before you start.
3 Look critically at the observations and measurements as you make them. Take them again if you are not satisfied. You may not be able to repeat them later.
4 Where possible, do repeats.
5 Include all units.

When making observations, try to be detailed.

For example, if you are carrying out experiments in a test tube you might look for:

- changes in colour
- changes in state, e.g. solid → liquid
- whether a gas evolved
- whether a precipitate formed
- a change in temperature – exothermic or endothermic
- cracking noises when heated.

When describing colour, try to qualify it. For example, yellow might be amber yellow, canary yellow, dark yellow, lemon yellow, etc.

Here is a table of results produced by a student for an investigation into the effect of temperature on the rate of reaction between sodium thiosulphate and dilute hydrochloric acid.

Temperature	1st	2nd	3rd	Average
40°C	20s	26s	22s	22.67s
50°C	18.5s	16.3s	16.5s	17.11s
60°C	13.7s	11.2s	12.2s	12.4s
70°C	9.8s	6.5s	4.7s	7s

At first sight this looks like a good table. However, there are a number of reasons why this table does not score O4b.

Why the table does not score O4b

1. There are only four temperatures considered and Health and Safety does not recommend doing this experiment above about 55°C (see page 83).
2. The units, although given, should be in the heading of the table.
3. Some results are taken to the nearest second while others are taken to 0.1 seconds and averages are given to 0.01 seconds. A hand-held stopwatch can give times to the nearest 0.1 seconds.
4. Were the starting temperatures exactly 40°C, 50°C, 60°C and 70°C? It would be difficult to measure such high temperatures accurately and maintain them.
5. Averaging 9.8s, 6.5s and 4.7s suggests the results are not very reliable.

How to get more than 6 marks

In order to score marks of 7 or 8, you should consider these three things.

1. You must have a large amount of data in a proper table with appropriate units, repeats and to the correct number of significant figures.
2. You need good quality data. If you have repeated the results three times, the three results should be close together.
3. It should be obvious that some skill was needed to get your results. Measuring the volume of gas collected in a measuring cylinder or timing with a stopwatch is unlikely to meet this requirement.

Example

Sally is asked to plan and carry out an experiment to find the concentration of ethanoic acid in a sample of vinegar.

She titrates a sample with 25 cm^3 of sodium hydroxide solution. Here are her results.

	Rough	1	2	3
Final burette reading (cm^3)	12.20	12.00	12.20	12.00
Initial reading (cm^3)	0.10	0.00	0.25	0.10
Volume of acid added (cm^3)	12.10	12.00	11.95	11.90

Average volume (ignoring Rough) = 11.95 cm^3

This table of results would be worth 8 marks.

Skill Area A

This Skill Area is about processing data, drawing conclusions and using scientific knowledge and understanding to analyse the conclusions.

On page 14 you will find information about drawing graphs.

The graph shows Tim's attempt to use Excel to plot the results obtained in the experiment on page 82.

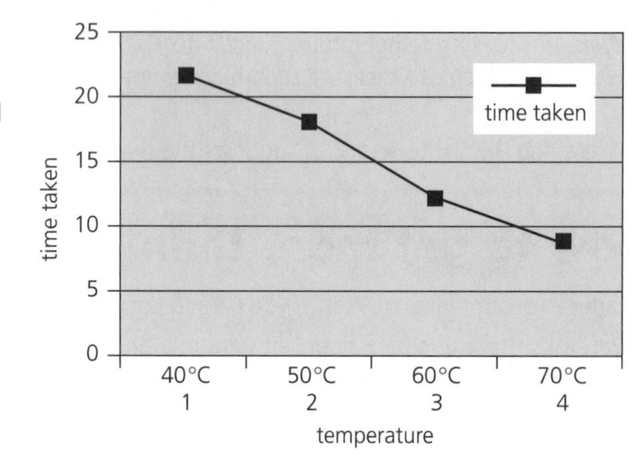

This graph does not score A.6a. The graph is not a scientific line graph but a disguised bar chart.

The correct line graph is shown below.

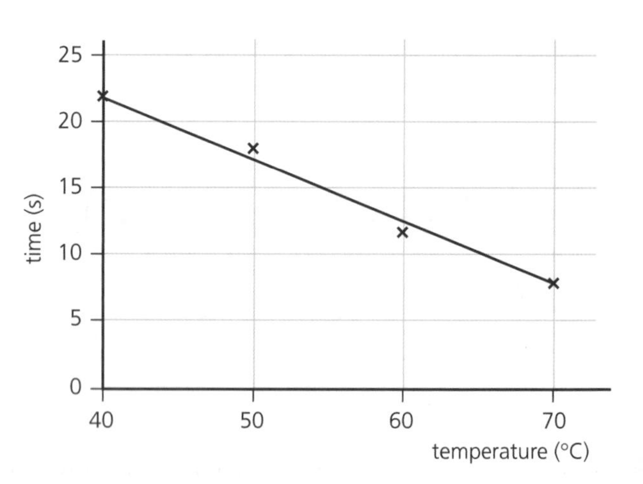

Plotting the graph correctly often shows a different trend in the results.

Sarah carries out an experiment testing the stretchiness of a nylon fishing line.

The diagram shows her apparatus.

The graph shows her results.

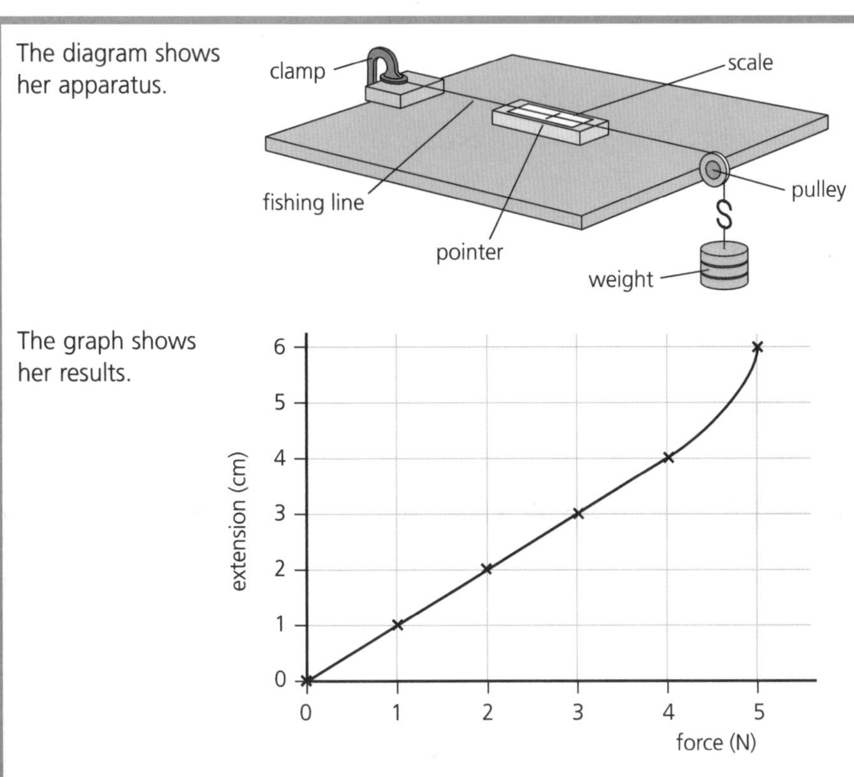

In her prediction Sarah stated that the extension of the nylon fishing line was directly proportional to the force. She explained this in terms of Hooke's law.

However, when she looks at her results she realises that her prediction was only correct in part.

Up to about 4N, the graph is a straight line through the origin. This means that the extension is directly proportional to the force over this range. However, above 4N the extension is no longer proportional to the applied force.

Sarah might now try to explain these results by considering the structure of a polymer such as nylon.

For a short extension, the chains remain intact so that the line returns to its original length when the force is removed. With larger forces the weak bonds between the chains are broken and the chains are extended. They cannot return to their original size when the force is removed.

The biggest fault students have with Skill Area A is not attempting to explain their results applying scientific knowledge and understanding.

Skill Area E

This Skill Area is about evaluating your investigation.

Evaluation requires you to do two things:

1 Consider the quality of the data collected.

- Is it sufficient?
- Are there any anomalous results?

2 Ask whether the investigation can be improved.

Students often suggest new investigations rather than improvements to the one they have done. For example, after studying the effect of concentration on a reaction, they suggest studying the effect of changing temperatures. This is not acceptable.

Sample investigation

Ali carries out an investigation into photosynthesis of pondweed. This is carried out using a lamp at different distances from the boiling tube.

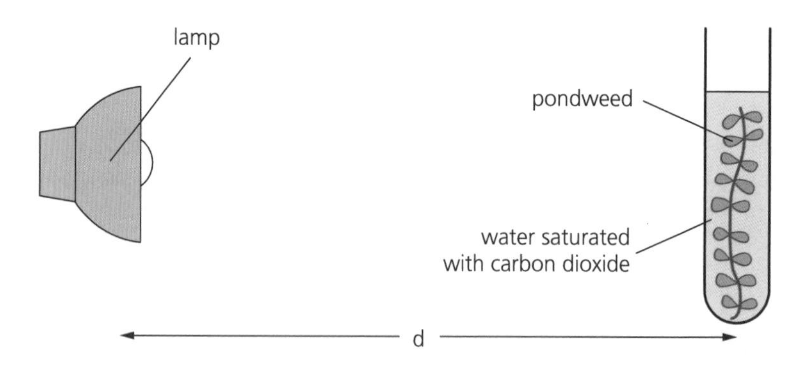

The number of bubbles produced every two minutes is counted.

Here are the results.

Distance in cm	Number of bubbles in 2 minutes			Average number
5	50	46	42	46
10	40	36	32	36
15	28	24	20	24
20	12	10	8	10

Ali plots a graph of number of bubbles every 2 minutes against $\frac{1}{d^2}$.

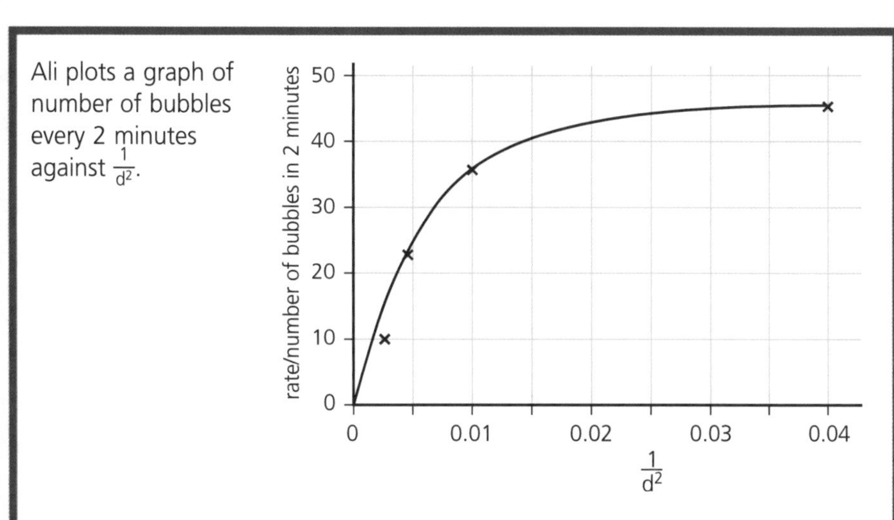

Ali had predicted that the rate of photosynthesis was proportional to light intensity. (Light intensity from a source is proportional to $\frac{1}{d^2}$.) If that were the case, the graph would be a straight line and this could be true for the first three results. The curve can be explained using limiting factors (page 55). To be secure we would need additional results for $\frac{1}{d^2}$ between 0.01 and 0.04.

Notice that the three results decrease each time. Is averaging therefore valid? Is this due to the presence of different amounts of carbon dioxide?

Ways to improve the experiment

1. Saturate the water in the boiling tube with carbon dioxide before each measurement.
2. Use a light meter as a better means of measuring light intensity.
3. Put a thermometer into the water to check that the temperature does not change when the lamp is close to the boiling tube.
4. Is there a better method than counting bubbles of oxygen gas? Bubbles can be different sizes. Only small volumes of gas are collected. Is there a way of measuring small volumes of gas?
5. The room is illuminated. How does this light affect the results? Should the investigation be carried out in a darkened room?

Ali does not have to try these improvements but could get additional credit for noting them. Avoid statements such as 'do it more often', 'do it more carefully' or 'use better apparatus'. These comments will get no credit.

Writing your report

The only evidence of your work after the event is your written report. This will be marked by your teacher and then possibly sent to an external moderator.

Writing the best possible report you can is very important. The report can be handwritten or done on a computer but it must be your own work.

If it is handwritten (and most reports are), write as clearly as possible. Your teacher may be used to your handwriting but this report will be read by other people who are not. It may also be photocopied and this does not improve the quality.

Useful tip
Write your report in four sections: Planning, Observation, Analysis and Evaluation. Use a highlighter pen to mark key points so they will not be missed by the examiner.

If your work is done on a computer, it must be done by you. Take care if using a spell checking programme: this can lead to mistakes. For example,

If you type in that a compound is 'ionically bonded', your spell checker will not recognise the term and will offer 'ironically' instead. Accepting this alternative will make your report nonsense.

There are no marks for spelling, punctuation and grammar but communication is built into the examination criteria. Read your report carefully before you finish to avoid simple mistakes.

The outline below and opposite will help you write your report. Make sure you use all of the headings in **bold** red type.

Title of investigation

1 Planning

What I think will happen.

This should include a prediction if possible.

Why I think this will happen.

This should include scientific knowledge you have researched from books, the Internet, etc. Don't copy out large bits of text. Put it in your own words. Include a reference that another reader can check.

List of the apparatus I plan to use.

In order to make the investigation fair, which factors:

- **affect how the investigation will work?**
- **will I keep the same?**
- **will I change?**

What I will count and what I will measure.

Decide on the range of measurements you will make, their frequency and the degree of accuracy you can expect from the apparatus you are using.

My detailed plan.

Any safety issues that will be important.

Preliminary work I have done.

2 Observation

Carrying out the investigation

Results should be in a table or chart.

Did I repeat my experiments?

3 Analysis

I used my results to produce a bar chart or a line graph (better) or a calculation involving complex maths.

What have I found out?

Are there any patterns in my results?

Do my results match with my prediction?

Can I explain what has happened, using scientific knowledge?

4 Evaluation

How reliable were my results?

Are there any results that do not fit the pattern?

If so, can I explain why these results were obtained?

If I were to repeat the experiment, what would I do to improve my results?

Glossary

a.c. generator p72–3 device that produces an alternating current

acceleration p68–9 the rate of change of velocity, measured in m/s^2

activation energy p24–5, p26–7, p28–9, p30–1 energy required to start a reaction

active site p32–3 area where interaction takes place in an enzyme molecule

active transport p20–1, p56–7 energy-requiring process that enables the movement of particles across a membrane, often against a concentration gradient

aerobic respiration p56–7 process in plants and animals that produces energy efficiently

alpha(α) radiation p38–9 a radioactive emission consisting of two neutrons and two electrons

ammeter p46–7, p48–9 device that measures the size of an electric current

anaerobic respiration p32–3, p56–7 the release of energy from a food molecule, such as glucose, in the absence of oxygen. *See fermentation*

anti-diuretic hormone (ADH) p66–7 hormone secreted by pituitary gland; associated with the regulation of water levels in the body

asexual reproduction p62–3 reproduction without sex, e.g. taking cuttings from a plant

atomic or proton number p34–5 number of protons in the nucleus of an atom

beta(β) radiation p38–9 a radioactive emission that consists of fast moving electrons

boiling p18–9 change in state from liquid to gas that takes place at the boiling point

brain p66–7 part of the nervous system in vertebrates continuous with the spinal cord. It coordinates information received and generates nerve impulses

calorimeter p60–1 a metal can in which experiments involving energy changes are carried out.

catalyst p30–1, p54–5 a substance that increases the rate of a reaction but is itself unchanged

catalytic converter p30–1 a device connected to a car's exhaust system to remove harmful pollutants such as oxides of nitrogen and carbon monoxide

centre of mass p70–1 point where total mass of body can be considered to lie

chain reaction p40–1 nuclear reaction that is self-sustaining

chemical formula p13 a shorthand way of representing a compound using symbols showing the numbers of different atoms combining

chlorophyll p54–5 green pigment in plants that is able to absorb light providing photosynthesis with a source of energy

chloroplast p54–5 the site for photosynthesis

chromosome p62–3 consists of a series of genes

close packing p44–5 particles, for example in a metal, that are as close together as is possible

compound p11 substance formed by joining atoms of different elements together

concentration p24–5 the quantity of solute dissolving in a volume of solvent, e.g. 10 g/dm^3

concentration gradient p20–1 measure of the difference in concentration between two places with respect to the same substance

condensing p18–9 changing of a vapour (or gas) into a liquid accompanied by the release of energy

conduction p44–5, p58–9 in electricity, where charged particles (electrons or ions) move through a material. In thermal conductivity, where energy is passed through a material

continuous variable p80–1 a variable where all values are possible, e.g. distance a car travels across a surface

convection p58–9 movement of parts of a fluid (gas or liquid) due to changes in density

covalent bond p36–7 formed by the sharing of a pair of electrons

crystalline structure p18–9 a structure with a regular shape due to a regular arrangement of particles within it

current p48–9 a flow of electric charge

deceleration (or negative acceleration) p68–9 rate of slowing down

delocalised electrons p44–5, p58–9 unfixed electrons that are free to move

denatured. p32–3 when the structure of a protein molecule is permanently changed, e.g. by high temperature or extreme pH

density p58–9 mass per unit volume of material, measured in kg/m^3 or g/cm^3

dependent variable p80–1 a variable that depends upon another. The volume of gas collected depends upon the time. Volume is the dependent variable

diabetes p66–7 condition caused by the failure of the body to control the concentration of glucose in the blood

diamond p44–5 form of carbon in which all the carbon atoms are strongly bonded into a three-dimensional structure

diffraction p52–3 the spreading out of a wave as it passes through a narrow gap or past an obstacle

diffusion p20–1, p22–3 spreading out of a substance to fill all the available space

diploid p62–3 nucleus containing two of each kind of chromosome

discrete variable p80–1 a variable that can only have certain values, e.g. shoe sizes

dispersion p52–3 the splitting of light into its constituent colours

dynamo p72–3 a device, e.g. on a bicycle, that converts kinetic energy into electrical energy

electromagnetic induction p72–3 a voltage is induced in a conductor when the magnetic field through it changes

electromagnetic radiation p38–9, p52–3 travels by transverse wave motion

electromagnetic spectrum p52–3 the range of electromagnetic radiation, namely gamma rays, X-rays, ultraviolet radiation, light, infra-red radiation, microwaves and radiowaves

electron p34–5 very small fundamental particle with a negative charge

electron shells p10, p34–5 where electrons are arranged in atoms

electrostatic forces p26–7 forces of attraction or repulsion between charges

element p9 pure substance made of one kind of atom

endothermic reaction p54–5 a reaction that takes in energy from the surroundings

energy level diagram p30–1, p62–3 diagram showing the energy content at stages during a reaction

enzyme p32–3 proteins that act as biological catalysts

equilibrium p22–3 a state of balance

evaporation p18–19, p58–9 process by which liquid changes to a gas at any temperature

exothermic reaction p30–1, p56–7, p62–3 reaction that gives out energy to the surroundings

fair testing p80–1 where all the variables except one are kept constant so any changes observed can be attributed to that variable

fermentation p32–3, p56–7 process by which enzymes in yeast convert glucose into carbon dioxide. Also anaerobic respiration

fission *see nuclear fission*

fertilisation p62–3 fusion of nuclei of two different gametes to form a zygote

freezing p18–9 the change of state from liquid to solid that takes place at the freezing point

frequency p24–5, p52–3 the number of oscillations or vibrations per second

Glossary

friction p70–1 force that opposes two surfaces from slipping and sliding over each other

fusion *see nuclear fusion*

gametes p62–3 a sex cell, e.g. sperm or egg cell

gamma(γ) radiation p38–9 a radioactive emission consisting of short wavelength electromagnetic radiation

Geiger counter p40–1 instrument for measuring radioactive decay with a Geiger-Muller tube

gene p64–5 a section of the DNA of a chromosome that either on its own, or with other genes, determines a particular characteristic

genetic engineering p64–5 technology that enables a gene (DNA) from one organism to be introduced into another organism

giant structure p44–5 crystal structure where all particles are linked into a network throughout the crystal, e.g. diamond

glycogen p66–7 how glucose is stored in the liver

graphite p44–5 one form of the element carbon. The carbon atoms are in layers with only weak forces between the layers

Haber process p30–1 industrial process for manufacturing ammonia from nitrogen and hydrogen using an iron catalyst

half-life p40–1 time taken for the number of undecayed nuclei in a sample of a radioactive isotope to halve

half-life curves p40–1 a graph of count rate *v* time that can be used to calculate half life

haploid p62–3 nucleus containing only one of each kind of chromosome

homeostasis p66–7 ability of a complex organism to maintain a stable internal environment for its cells and tissues

Hooke's law p86–7 the extension of a spring is directly proportional to the force applied

hormone p66–7 chemical secreted into the bloodstream to act as a chemical messenger

hypothalamus p66–7 part of the brain

independent variable p26–7, p82–3

insulator p42–3, p58–9 substance that does not allow heat energy (thermal insulator) or electricity (electrical insulator) to pass through it

insulin p66–7 hormone secreted by cells in the pancreas to control blood sugar levels

ion p26–7, p34–5, p42–3 charged atom or group of atoms

ionic bonding p26–7 bonding between metal and non-metal atoms involving the complete transfer of one or more electrons from metal to non-metal and the formation of ions

isotope p38–9 atoms of the same element but containing different numbers of neutrons

kinetic energy p70–1 energy an object has due to its movement

lactic acid p56–7 product of anaerobic respiration

light gate p68–9, p70–1 a device linked to a computer that can be used for accurate timing. The timed object cuts a beam of light that starts or stops the clock

limiting factor p54–5, p90–1 factor that holds back a process, e.g. low light intensity can limit photosynthesis

liver p66–7 largest human organ; has many functions including producing bile and urea and storing glycogen

lock and key model p32–3 model used to explain how enzymes work

magnetic field p52–3 a region around a permanent magnet or electric current where forces are exerted on magnetic materials

main sequence p42–3 phase in the life of a star when it is generating energy by nuclear fusion of hydrogen

mass or nucleon number p34–5 number of protons and neutrons in an atom

meiosis p62–3 type of cell division producing cells that have half the number of chromosomes of their parent cell

melting p18–9 a solid changes to a liquid at the melting point

mitosis p62–3 type of cell division that produces identical copies of cells

National Grid p72–3 system for transporting electricity around the country at very high voltage

negative feedback p66–7 by producing more of something, the source causes its own shutdown

neutron p34–5 neutral particle found in the nuclei of atoms

neutron star or black hole p42–3 compact object consisting of densely-packed neutrons

nuclear fission p40–1 the splitting of an atomic nucleus into fragments

nuclear fusion p40–1, p42–3 fusing together nuclei of light atoms to form a heavier atom with the release of large amounts of energy

nucleon p34–5 proton or neutron

nucleus p34–5 centre of an atom where protons and neutrons are found

Ohm's law p48–9 law concerning the relationship between current, voltage and resistance

osmoregulation p66–7 ability of animals to regulate water content

osmosis p22–3 movement of water from a solution with a high water concentration to one with a lower water concentration through a selectively permeable membrane

ovary p62–3 female sex organ; produces eggs or ova

oxygen debt p56–7 when a cell can no longer respire aerobically and has to respire anaerobically

pancreas p66–7 organ that secretes insulin

parallel p46–7, p48–9 circuit or part of a circuit where there is more than one current path

partially permeable membrane p22–3 membrane that allows water molecules to pass through but not larger molecules such as glucose

pendulum p70–1 swinging or oscillating weight suspended from a string or rod

period p8 horizontal row in the Periodic Table

Periodic Table p8, p34–5 classification of elements in order of atomic numbers. Elements with similar properties are found in columns called groups

pH p32–3 measure of acidity or alkalinity on a numerical scale

photosynthesis p20–1, p54–5 process where green plants use sunlight to manufacture glucose

pituitary gland p66–7 small gland connected to hypothalamus at the base of the brain

pivot p70–1 point about which an object turns

polyethyne p44–5 synthetic polymer formed from ethyne; a good conductor of electricity

polymer p44–7 material made up of long chains of shorter molecules called monomers

potential energy p70–1 energy of an object based on its position

precipitation p50–1 a reaction in which a solid is formed when two solutions are mixed

prediction p80–1 use of scientific knowledge to make a forecast

protein p32–3 large molecule that is a polymer of amino acids

proton p34–5 positively-charged particle in the nucleus of an atom

radiation p58–9 given out when an unstable nucleus changes to a more stable one; also used to describe any electromagnetic wave given out by an object

radioactivity p38–9 the breakdown of an unstable nucleus, leading to the emission of particles or electromagnetic radiation

rate of reaction p24–5 the speed with which products are formed or reactants used up during a chemical reaction

red giant p42–3 star that has gone through its main sequence, expanded and cooled

reflection p52–3 change in direction when light or other wave motion rebounds at a surface between two materials

refraction p52–3 change in direction when light or other wave motion passes from one material to another

resistance p48–9 measure of the opposition of a conductor to electric current passing in it

respiration p20–1, p56–7 respiration is a process in plants and animals where cells release energy

root hair cells p22–3 specialised cells in plant roots with a large surface area allowing water to enter the plant

selective breeding p64–5 artificial selection by humans to improve animal breeds, crops, etc.

series p46–7, p48–9 circuit or part of a circuit where there is only one current path

skin p66–7 extensive organ covering most of the surface of a body of a vertebrate

specific heat capacity p60–1 energy required to raise the temperature of one gram of material through 1°C

star p42–3 a large mass that gives out light due to energy generated in nuclear fusion reactions

step-down transformer p52–3 transformer used to decrease voltage

step-up transformer p52–3 transformer used to increase voltage

stomata p20–1 holes in the underside of a leaf where gaseous exchange takes place

subliming p18–9 direct change from vapour to solid without going through the liquid state. Also used to describe the reverse process

supergiant p42–3 very large, extremely luminous star

supernova p42–3 exploding star

surface area p28–9 the area of the surface of a solid object, usually measured in cm^2

testes p62–3 male sex organ where sperm cells are produced

transformer p72–3 electromagnetic device that changes the size of an alternating voltage

transverse waves p52–3 vibrations are at right angles to the motion of the wave, e.g. light.

vacuum p58–9 a region of space that contains no matter

variable p80–1 a factor that can be changed

variable resistor p48–9 a device for altering the current in a circuit by changing the resistance

velocity p68–9 speed in a definite direction

visible spectrum p52–3 colours of light visible, for example, in a rainbow: red, orange, yellow, green, blue, indigo, violet

voltage p46–7, p48–9, p52–3 measure of the difference in potential between two points in a conductor when a current passes

voltmeter p46–7, p48–9 instrument for measuring voltage

wavelength p52–3 length of one complete cycle of wave motion

white dwarf p42–3 star produced when a red giant contracts due to gravitational forces

xylem p22–3 plant tissue used in the passage of water

zygote p62–3 cell formed on the fusion of two gametes